D0081683

OTHER BOOKS BY MURRAY LORING

Risks and Rights of Animal Ownership
(Arco Publishing Company, Inc., New York)

Your Horse and the Law
(Cordovan Corporation, Houston)

OTHER DADANT PUBLICATIONS

American Bee Journal
American Honey Plants
Beekeeping Questions and Answers
Contemporary Queen Rearing
First Lessons in Beekeeping
Honey in the Comb
Instrumental Insemination of Honey Bee Queens
The Hive and the Honey Bee
The Honey Kitchen

BEES

and the

LAW

MURRAY LORING DVM, JD

FIRST EDITION

DADANT & SONS • HAMILTON, ILLINOIS

Second Printing 1982
Third Printing 1984

Library of Congress Catalog Card Number 80-66362
ISBN Number 0-915698-07-2

PRINTED IN THE U.S.A. BY JOURNAL PRINTING COMPANY
CARTHAGE, ILLINOIS

TABLE OF CONTENTS

IN GRATITUDE

This book could not have been written without the collaboration of my dedicated wife, Mildred.

Murray Loring

INTRODUCTION

It is ironic that at a time when the number of bees and beekeepers are increasing at the highest rate since the turn of the century, the simple pleasure of keeping bees is becoming increasingly difficult.

It is perhaps a sign of the times that an individual must decide at the outset of a project whether or not his actions will infringe upon the rights of others and weigh the consequences.

Beekeeping is no exception. There are people who have no use for bees, do not like them and do not wish to be around them. Although such an attitude is often based on a lack of knowledge about bees and beekeeping, these people do have a right to feel that way, and according to our system, their rights must be protected.

The nature of the bee dictates this philosophy. Although the activities and behavior of the honey bee have been the subject of intense study and manipulation for hundreds of years, the bee remains untamed and free, retaining the prerogative to sting in self-defense. If an innocent bystander is injured, or property is damaged, the beekeeper is often held liable.

Murray Loring DVM, JD, recognizes this fact and has experience to reinforce his opinion because, not only is he an attorney, he is an accredited veterinarian.

Dr. Loring became aquainted with THE AMERICAN BEE JOURNAL writing numerous articles. It was only a matter of time before he was asked to put his expertise to good use. The result is this fine effort.

The numerous citations included are intended for the attorney or apiary enthusiast with an interest in the law and bees. They will be invaluable as quick references to cases Dr. Loring has already researched.

Dr. Loring has written this book with the layman in mind, but he did have the foresight to include information an attorney and apiarists need to build a sound case for beekeeping.

Howard Veatch
DADANT & SONS, INC.

ACKNOWLEDGMENT

It is the scope of this book to bring together the law of the English speaking world which relates to the honey bee. To this end the reports of all courts of the last resort have been examined for decisions affecting her status or the rights and liabilities of her owners as affected by the fact of her ownership. The volumes which contain the acts of parliaments, congresses, and legislatures have also been examined for legislative acts concerning her. The quantity of matter available which concerns so small a morsel of the animal kingdom is staggeringly large.

In the drafting of this book, we have drawn from all sources, including an earlier work on the subject entitled "A Treatise on the Law Pertaining to the Honeybee" by the legal department of The American Honey Producers' League published by The American Honey Producers' League, Madison, Wisconsin, in cooperation with The American Bee Journal and Gleanings in Bee Culture in 1924. Although the book has long been out-of-print, sections have been reproduced here because they still hold important truths for the beekeeper and his legal positions.

The court cases discussed generally and specifically in this book are, in some cases, the stepchildren of earlier verdicts and have served as precedents for later decisions. If the discussions may serve as introduction to the legalities concerning the honey bee and if the gathering of the court cases may serve as a starting point of legal research in times of difficulty, the effort will have been well served.

PREFACE

DEFINITIONS

When the layman approaches a legal situation, he quite often finds himself at a distinct disadvantage. He may have spent a lifetime laboring under the delusion that he was a law-abiding citizen (as most citizens are,) and then find at some unexpected moment that he and his activities are the subject of litigation. Few people, outside the legal profession, are cognizant of the host of legalities that may be brought to bear upon them and their activities at any given time, and fewer are adequately prepared to deal with them.

To this average person it may seem, in times of such difficulties, that one need only determine what the law is and then comply with it, but the teeming courts are witness that this is often not the case. To compound the problem, the law, even when written, often appears conflicting and confusing; the language of the law may seem needlessly complicated and medieval in construction; and the very atmosphere of the courts present a sinister and threatening visage to the bewildered individual.

Man has often found himself in legal difficulties because of his relationships with animals. Dogs bite, horses kick and bees sting. Now if it is your dog or horse or bee and you received the bite, kick or sting, the matter usually lies directly between you and your animal, although laws exist to temper even those relations gone wrong. Introduce the third party into this potentially volatile situation and the possibility of legal action becomes very real.

The average person, and this includes many lawyers, regards the law as an exact science; a set of rules governing our relations with our government and with our neighbors. This view is perhaps correct in theory and in the abstract. But in actual practice, and in the concrete case, it is not true.

The law is a system of philosophy distilled from different cases occurring at different times in different places. Its aim is to regulate the relations of the individual with his government and with his fellow beings, in accordance with those principles. The ideal of the law is that the application of those principles to the particular situation will, in the view of the community at the time, accomplish justice. If justice fails in a particular controversy, the failure is due to the limitations of humanity. Such a failure may manifest itself in a bad law enacted by a short-sighted legislative body, the narrow view of the particular authority administering the law, or in the defective vision of the persons declaring that justice has failed.

The law is a science by means of which mankind strives to preserve order and make communities, cities, states and governments possible and relations among men tolerable. It is not an exact science like chemistry or mathematics as it always involves a proportion, one mean and one extreme of which contain the human element. As this human element varies with individual temperament, desires, purposes and educations and shifts with the ebb and flow of the tide of civilization, so does the law and its interpretations. Times and attitudes change. For example, a rural community may have seasoned its interpretation of justice with agricultural overtones for years and then find, as urban communities creep outward from city boundaries, that attitudes toward justice have changed with the times and the landscape. The farmer's manure pile, once a fixture of every farm, is now an eyesore to the urbanized community and a likely target for litigation. Livestock (including bees) and their habits are no longer as welcome or tolerated as they once were. A few years earlier, a controversy that would probably have decided in favor of an agricultural pursuit may now find the scales of justice tipped in favor of the resident who wants his patio free of the barnyard smells and/or visits from cows, sheep, or bees.

In the exact sciences such as mathematics or chemistry or astronomy or physics, the application of a given rule to the concrete case always produces the same definite result. The saving grace of the law and our method of its interpretation is its flexibility. In considering the law in any of its relations to the problems of humanity, we are dealing with a set of circumstances to which the standards of justice of the moment have been applied, a conclusion reached, and that conclusion transmitted for our guidance under similar conditions.

The central purpose is the doing of justice. We interpret justice to be the application of principles which satisfy the reason and the conscience of mankind. We have distilled these principles from the ideals of the majority of the community and they are interpreted for the moment by those whose duty it is to decide the controversy. Aside from the human element undertaking to do absolute justice, many controversies are complicated by a failure in presentation.

Rights depend upon evidence, and if this is not presented, or interpreted, in such a manner as to demonstrate the right and make clear the rule of justice, the wrong party may be cast. In our field this presentation may fail because of a lack of special knowledge on the part of the court, jury and counsel, of matters which to every beekeeper are elementary and which the parties have failed to supply.

SOURCES OF THE LAW

It is not so much to know the law as to know where to find it. So one of the important things is to know the sources of the law. Many persons look upon law as a set of rules passed by some legislative body and they are right as to part of the law only. A very small part of it is made up of legislative enactments, but by far the larger part is in those decisions of earlier controversies which all the courts of the English speaking world regard as guideposts in the search for justice in smaller controversies.

To speak simply, law has come down to us in layers. The first law was undoubtedly ancient custom. The Latin works of Justinian and other ancient writers expanded and codified the unwritten rules of ancient custom. In our own particular inheritance of law, we must also consider the lineal descent through Great Britain and the Acts of Parliament, the Acts of Parliament organizing the provincial governments, the acts of provincial assemblies and the decisions of the then contemporary courts of law. Add to this the Constitution of the United States and of the several states, the Acts of Congress and of several states and the decisions of the federal, state and local courts of law. Additionally, in all these courts of the English speaking world, the decisions of every other court in the contemporary world are read with respect and regarded as persuasive authority.

When we contemplate all this, that this body of law is contained in tens of thousands of volumes, and that, in addition, the law of portions of our own country and the British Provinces is tinctured with Spanish and French law, we realize the impossibility of one knowing it all, and of greater importance of knowing where to find it.

RELATION OF THE LAW CONCERNING BEEKEEPING TO OTHER BRANCHES

When we remember the definition of the law, regard its central principle and purpose, and bear in mind its sources, we are impressed with the impossibility of dividing the field so as to consider by itself, the law pertaining to beekeeping.

Lawyers, judges and juries apply our inheritance of law in that they reason by analogy. They apply to controversies, principles announced in or decided by analogous cases. Hence a decision concerning a horse, a cow, a dog, a bird, may by parity of reasoning be made to fit the case of the bee. Title to a colony of bees may be decided by reference to a decision as to the ownership of a domesticated deer, or the ownership of a new swarm may depend upon what some other court has decided as to the ownership of the fawn of the same deer.

Liability for injury by bees may depend on the principles applied where someone was injured by a ferocious dog or a mad bull, and this upon the negligence of the owner, and whether the person injured was negligent, and whether that negligence was the proximate cause of his injury. So whether an apiary may be kept in a certain place depends upon whether it is a nuisance, and may in turn depend on whether noisy fowls, a barking dog, a filthy pig, or a noisome stable has been held under similar circumstances to be a nuisance.

Lawyers and judges are always looking for the decision in an analogous case, a precedent as it is called, and this analogous case may not, because of the beekeeper's special knowledge, appear to him to be parallel at all. It is the duty of his counsel to demonstrate that special knowledge to the court, so that he may not be cast by the application of an inapplicable analogy.

APPLICATION OF THE LAW—WHAT TO DO IN TROUBLE

An interest in legal difficulties usually indicates that it is too late for the old adage, "Avoid trouble!" However, for the beekeeper, the initial step in the process of a successful litigation is usually just that. Familiarize yourself with local, state and federal legalities touching upon your business and obey them. Avoid trouble and be the best beekeeper you know how to be, even if it means placing some extra effort into public relations. However much beekeepers may love honey bees, every individual has his or her own memory of being stung and the general public attitude is not likely to be in favor of the beekeeper unless he has done some promotional work in that direction.

Once in trouble, compose your differences. Litigation is expensive and an economic waste in non-productive effort. A reasonable compromise is usually better than the judgment of the court. Matters of small amount might better be forgotten than litigated.

Consider, too, that the decision of the court, for good or ill, will have an effect upon the whole beekeeping industry. Court cases are favorite topics of conversation. Facts and emotions of the case are often discussed out of context or just as often repeated incorrectly by those who have no firsthand knowledge. It would be foolhardy to collect a few dollars in damages only to find the public image of beekeeping has been injured to the point of loss of future business or subsequent adoption of a prohibitive ordinance.

If all efforts at settlement fail and the principle at stake or the amount involved warrant and you must litigate, put yourself in the place of the other fellow and ask yourself, "Am I in the right? Is my cause just?" If the questions are answered in the positive, it is best to consult a lawyer familiar with the statutes and ordinances of your locality. The law does vary somewhat from place to place, from time to time.

Having selected a lawyer, follow his advice and counsel and if he says to litigate, see the case through to the court of last resort using every fair and honest means to win.

If you are defeated, avoid the generalities of laying the blame upon the court, the counsel and witnesses. Such generalities do no good and may do a great deal of harm. The courts are honest, the great majority of lawyers doubtless do the best they know, and 99 per cent of the cases are fairly and correctly decided. Consider that it is possible that the defeat could be laid directly or indirectly to your own narrow and partisan view of the circumstances and your own mistaken notions of justice.

If you win, temper justice with mercy and do not let the gloat of victory dull your sense of honesty, fair dealing and humanity. Remember that the goal of being in business is to operate your business quietly, efficiently, and profitably, generating public good will along the way to increase those profits. Consider the notoriety of the litigation as a wound to be healed as quickly as possible.

THE STATUS OF THE HONEY BEE IN THE LAW

The law divides the entire animal kingdom into two classes:[1]

First, those which are domesticated (ferae domitia) and, second, those which are wild (ferae naturae). The rights and liabilities of persons with reference to the animal kingdom then are likewise divisible. Bees belong with the latter class and, in considering the law with reference to these cases, rules pertaining or applicable to the former class would not have any significance. Wild animals are also divisible into two classes:

Those which are free to roam at will, and those which have been subjected to man's dominion. Rights and liabilities depend upon the class into which the animal falls at the particular time. If it be in a State of Nature, free to roam at will, it is the property of no one, not even of the one on whose land it may be at the particular time, and may become the property of the first taker even though he be a trespasser and liable for the trespass.

One who enters another's premises without the invitation or permission of the owner is a trespasser, but this gives the owner of the premises no title to the wild things thereon, it merely gives him the right to protect others from coming thereon and taking them. If, however, some person against his will enters the premises and takes a wild animal or a swarm of wild bees, such a person becomes the owner of what he takes, but he has to answer to the owner of the premises for the trespass. If, however, the animals have been brought within the dominion of the owner of the premises as deer in a park, rabbits in a warren, or bees in a hive, such an entry and taking would be a crime as the law recognizes the property of him who has dominion over them and the taker would gain no title by the taking, for the owner might regain them by legal proceedings.[2]

So we may understand that the animal kingdom to which the bees belong is subject to a certain qualified proprietary interest. That is, they belong to no one, not even the owner of the soil on which their nest may be unless they have been subjected to his dominion, and when so reduced to possession, they are his property. This principle, however, is subject to an important modification: they remain the property of the possessor only so long as his dominion continues, and if such animals regain their freedom, as bees by swarming out and occupying some natural hive, as a hollow in a tree, the property right is lost and they again revert to their natural state and become the property of the first taker.

These same notions also control the matter of liability for injuries done by the bees, and such liability depends on proprietorship. So it would seem that if the bees have escaped from their owner, or have swarmed out of his

1. Blackstone Commentaries, Book II, p. 390.
2. Blackstone Commentaries, Book II, p. 392.

hive, unless he can be shown negligent in having permitted this, there can be no liability for injuries done by them.

It is to be understood that any attempt to correlate "legalese" to the words of the layperson may have obvious pitfalls. Oversimplification becomes inevitable.

Fig. 1. Backyard beekeeping means that every beekeeper must practice good public relations as well as good beekeeping. This hive is located away from patios, play areas, swimming pools, confined pets, and neighboring doorways and driveways. The shrubbery behind the hive will direct the flight of the bees at least seven feet above the ground at boundary lines.

CHAPTER I

STATUTES, ORDINANCES AND REGULATIONS

A land flowing with milk and honey
. . . Old Testament

A discussion of *statutes, ordinances* and *regulations* will be more intelligible if the terms connected with this chapter are defined.

Statute: an act of the Legislature declaring, commanding, or prohibiting something.[1]

or

a rule of action which is prescribed by the corporate power in a state and which all persons within sphere of its operation, are compelled to obey.[2]

or

is the written will of the Legislature, solemnly expressed according to the form necessary to constitute it a law of the state.[3]

or

is simply a fresh particle of legal matter dropped into the previously existing ocean of law.[4]

Ordinance: is a local law, a rule of conduct prospective in its operation, applying to persons and things subject to local jurisdiction.[5]

or

The word "ordinance," as applicable to the action of a municipal corporation, should be deemed to mean local laws passed by the governing body. A municipal corporation passes laws called "ordinances," and enacts rules.[6]

or

is a local law and binds persons within the jurisdiction of the municipal corporation.[7]

Regulation: Any rule for the ordering of affairs, public or private, whether by statute, ordinance, or resolution.[8]

or

The word "regulation," "rule," "ordinance," and "by-law" are synonymous terms.[9]

The enforcements in the title of this chapter have been recorded for decades, perhaps centuries, in one form or another by, for and against bees and bee owners. Whereas the records are not available as to the earliest, an ordinance enacted in the 1880's, or prior, can be a starting point. In Arkansas, the City of Arkadelphia passed the following:[10]

"Be it ordained by the City Council of the City of Arkadelphia, that it shall be unlawful for any person or persons to own, keep or raise bees in the City of Arkadelphia, the same having been declared a nuisance; that any person or persons keeping or owning bees in the City of Arkadelphia are hereby notified to remove the same from the corporate limits of the City of Arkadelphia within thirty days from date hereof."

Incidentally, the Arkansas Supreme Court declared this ordinance invalid, in 1889, as being too broad.

At least since the 1880's, literally hundreds of various types of enforcements have been enacted. Some have been repealed, some said to be void, and in some instances, declared unconstitutional. However, many remain on the books of cities and towns, various states, and the Federal government. They cover a variety of bee legislation including, but not limited to, disease, taxation, transportation, pesticides, zoning, et cetera, et cetera. As an example, an ordinance on the books of the City of Los Angeles, reads (in part):[11]

> "No person shall keep any bees in or upon any premises in this City. Nothing in this section shall be deemed or construed to prohibit the keeping of bees in a hive or box located and kept within a school house for the purpose of study or observation, or to prohibit the keeping of bees within that certain territory annexed to this city on May 22, 1915, and known as the 'San Fernando District,' and that certain territory annexed to this City on June 14, 1916, and known as the 'Westgate District'."

In Michigan, a statute was designed to prevent the spread of bee diseases by prohibiting bees being brought into the State on combs, used hives, and other used apiary appliances. The statute was contested. In a 1954 ruling it was declared to be in the interest of public health and did not constitute an unconstitutional burden upon interstate commerce. Two other states have similar statutes as the one in Michigan. The state enactment reads (in part):[12]

> "It shall be unlawful for any person, firm, corporation or transportation company to bring into this State any bees on combs, used hives, or other used apiary appliances from any other states or other countries; provided however, that common carriers may transport bees and apiary appliances through this State if the shipment originated outside of this State and is destined for some point outside of it . . . . In addition to the penalties herein before provided, bees on combs, used hives or other used apiary appliances brought into this State in violation of the provisions of this act shall be confiscated and destroyed."

A Florida statute, requiring an application for a permit, to transport bees into the State, be accompanied by a certificate certifying that the bees have been inspected within a period of 30 days . . . ., was challenged by a resident, and migratory beekeeper, of Florida. He contended the certificate requirements were unnecessary, onerous, burdensome and unconstitutional. The statute is as follows:

> "All honey bees (except bees in combless packages) and used beekeeping equipment shipped or moved into the State, shall be accompanied by a permit issued by the commissioner. Before any bees (except bees in combless packages) or used beekeeping equipment is shipped or moved from any other state into the State, the owner thereof shall make application on forms provided by the commissioner for a permit. The application shall be accompanied by a certificate of inspection signed by the state entomologist, state apiary inspector or corresponding official of the state from which such bees or equipment are shipped or moved. Such certificate shall certify that all

of the colonies, apiaries, and bee yards, owned or operated by the applicant, his agents or representatives, have been inspected annually at a time when the bees are actively rearing brood, including one inspection within the period of thirty days immediately preceding the date of shipment or movement into Florida and that no American foulbrood or other contagious or infectious diseases have been found in any colony, apiary, beeyard, or other places where bees or equipment have been held by the applicant, within the period of two years immediately preceding the date of shipment or movement into Florida; provided that when honeybees are to be shipped into this State from other states or countries wherein no official apiary inspector or state entomologist is available, the commissioner may issue permit for such shipment upon presentation of suitable evidence showing such bees to be free from disease."

After judicial study, the United States District Court's ruling was adverse to the migratory beekeeper. The statute, they ruled, did not impose an unconstitutional burden, did not unduly burden interstate commerce and was reasonably adapted to the end sought.[13]

To this point, the discussion, somewhat brief, has exemplified statutes and ordinances and mentioned regulation. All these enactments are, in a sense, legalities. That is, they are commonly termed "laws."

A Federal court, taking note of the terms of this chapter, defined law as:

> The whole body of rules of conduct applied and enforced under the authority of established government in determining that which is proper and should be permitted and that which should be denied, or even penalized, in respect of the relation between a person and the state, between him and society, or between him and another individual, including a provision of a constitution, a legislative enactment or statute, a municipal ordinance, a principle declared in an authoritative decision of a court, a rule of practice prescribed by a legislature or promulgated by a court acting with authority, even, to some extent, a usage or custom.[14]

In matters concerning bee diseases occurring in, or suspected of prevailing, in foreign countries, the Federal Government invariably enacts laws restricting bee imports from these countries. Acarine disease, sometimes called Isle of Wight disease, is a prime example.

Federal laws have been passed to prohibit importation of honey bees into the United States from all countries but Canada. Under the existing law importations cannot be made by the United States Department of Agriculture for its own experimental purposes, not even from countries in which acarine disease does not exist.[15]

Conversely, the Federal Government has no laws regulating the interstate movement of honey bees, but the Honey Act of 1922, revised in 1947, 1962, and 1976, prohibits most importation of all stages, germ plasm and semen of honey bees into this country. An exception is the provision in the Act that provides for the importation of honey bees from countries which the Secretary of Agriculture has determined are free from harmful diseases, parasites and undesirable species or subspecies of honey bees. Also, the United States Department of Agriculture authorizes the unrestricted importation of all stages of the honey bee from Canada. Even importations by the United States Department of Agriculture for experimental purposes require a permit from the Animal Plant Health Inspection Service.

Laws regulating inter- and intrastate movement of honey bees are the responsibilities of the various state legislatures and, as such, do vary from state to state. In the following pages, "Michael's Summary (1979) of Bee Disease Laws of the United States" is presented for beekeepers and apiary inspectors. However, it is recommended that apiarists, interested in the inter- or intrastate transportation of bees and/or equipment, contact the proper authorities for the latest particulars on entrance requirements, permits, laws (local or state) and regulations. Copies of laws and regulations are available from the various state departments of agriculture.[16]

The Honorable Judge in the Florida case, mentioned heretobefore, added a bit of legal philosophy that is appropriate, not only for this chapter, but for all facets of the jurisprudence system.

> "In this land of liberty under law in which we have the good fortune to live, any law, because it circumstances or affects, at least to some degree, freedom of action of individuals, may to that extent be considered or viewed as a hardship and onerous upon individuals subject to it. Yet we would have it no other way because we know that, without our government of ordered liberty under law, we may loose all our precious freedom. The migratory beekeepers in Florida have recognized and accepted, that fundamental principle."[17]

The gamut of "bee laws" — statutes, ordinances, regulations, court decisions (those that leave their legal imprint) — will be discussed in their varied criteria throughout the chapters of this book.

CITATIONS

1) — *Young v. Gerosa,* 202 N.Y.S.2d.470,478, 11 A.D.2d.67
2) — *Drugan v Fialer,* 139 N.E.2d.704, 705
3) — *Lane v. Missouri County Comm'rs.,* 13 P.136, 139, 6 Mont.473
4) — *State v. Rechnitz,* 52 P.264, 265, 20 Mont.488
5) — *C.I.R. v. Schnackenberg,* C.C.A., 90 F.2d.175, 176
6) — *Armatage v. Fisher,* 26 N.Y.S.364, 367
7) — *Pennsylvania Co. v. Stegemeier,* 20 N.E.843, 118 Ind.305
8) — *Kepner v. Commonwealth,* 40 Pa.St.124, 129
9) — *State ex rel. Krebs v. Hoctor,* 120 N.W.199, 200, 83 Neb. 690
10) — *City of Arkadelphia v. Clark,* 11 S.E.957, 52 Ark.23
11) — *Ex Parte Willis,* 81 P.911, 11 Cal.2d.571
12) — *Wyant v. Figy,* 66 N.W.2d.240, 340 Mich.602
13) — *Trescott v. Conner,* N.D. Fla., 390 F.S.765
14) — *Strother v. Lucas (US)* 12 Pet.410, 9 L.Ed.1137
15) — *Eckert and Shaw* — "Beekeeping" The Macmillan Company
16) — *H. Shimanuki* — American Bee Journal, March 1980
17) — *Id,* N.D. Fla., 390 F.S.765

MICHAEL'S 1979 SUMMARY OF BEE DISEASE LAWS OF THE UNITED STATES

Compiled By

H. Shimanuki, Bioenvironmental Bee Laboratory
Agricultural Research Science and Education
Administration, USDA, Beltsville, MD, 20705

GLOSSARY

Certificate — a point of origin report issued by apiary inspector after inspecting honey bee colonies.

Drug Treatment — (a) Control — Feeding of chemotherapeutic agents after a disease has been detected. (b) Prevention — Feeding of chemotherapeutic agents to prevent appearance of disease symptoms.

Food Restriction — no honey should be used in preparing feed for package bees and queens in their respective shipping cages.

Permit — point of destination document permitting the movement of honey bee colonies and/or equipment inter- or intrastate.

Location Controlled — restrictions on the establishment of apiary sites.

TABLE 1 — U. S. INTRASTATE LAWS

	Registration of			Inspection of		Inspector			
State	Apiary Required	Queen Apiary Required	Identification of Apiary Required	Apiary Required	Honey House Required	Has Right of Entry	Must Disinfect Equipment	Inspection Certificate Required	Permit for Movement of Bees and Equipment Required
Ala.	x		x	x		x		x	x
Alaska									
Ariz.	x		x	x		x		x	x
Ark.	x		x	x		x	x	x	x
Calif. (1)	x		x			x			
Colo. (2)			x			x			
Conn.	x		x			x		x	x
Del.	x	x		x		x	x	x	
Fla.			x	x	x	x		x	x
Ga.	x	x		x		x		x	
Hawaii									
Idaho	x	x	x	x		x	x		x
Ill.	x	x	x	x		x		x	x
Ind.	x			x		x			
Iowa	x			x		x	x		
Kans.	x			x		x		x	x
Ky.	x			x		x		x	x
La.	x			x		x		x	x
Maine	x			x		x		x	
Mass.				x		x		x	x
Md.	x	x		x	x	x			
Mich.	x	x				x	x		x
Minn.	x			x	x	x	x	x	x
Miss.			x	x		x		x	x

Mo.						x			
Mont.	x	x	x	x		x			
N.C.		x		x		x			
N. Dak.	x		x	x	x	x	x	x	x
Neb. (1)	x	x	x	x	x	x		x	x
Nev.	x	x	x	x		x	x		x
N.H.				x		x			
N.J.				x		x		x	
N. Mex.	x			x		x		x	
N.Y.	x			x		x			x
Ohio	x	x	x	x		x		x	x
Okla.	x	x				x	x		
Oreg.	x		x	x	x	x	x	x	x
Pa. (1)				x		x	x	x	
R.I.	x	x	x	x		x	x		x
S.C.					x			x	x
S. Dak.	x			x	x	x	x	x	x
Tenn.	x			x	x	x	x	x	x
Tex.		x	x	x		x		x	x
Utah (1)	x		x	x	x	x	x	x	x
Va.				x	x	x	x	x	x
Vt.	x	x		x		x	x		x
Wash. (1)	x	x	x			x	x		
Wis.					x	x			
W. Va.	x	x	x	x	x	x		x	x
Wyo.	x		x	x	x	x	x	x	x

(1) Wax Salvage plant permitted

(2) Beekeepers with 25 hives or under are exempt from **all** provisions of the law unless they declare they are selling honey.

TABLE 2 — U. S. INTRASTATE LAWS

	Owner				AFB Diseased Colonies or Equipment					Drug Treatment		Hives
		Subject to										
State	Must be Notified	Tax	Fees	Penalties	Must be Quar-antined	Must not be Con-cealed	Must not be Exposed	Must not be sold or Transferred	Must be De-stroyed	Permitted for Control	Permitted for Prevention	Must have Movable Frames
Ala.	x		x	x	x	x	x	x	x		x	x
Alaska												
Ariz.	x				x		x		x	x	x	x
Ark.	x			x	x	x	x	x	x		x	x
Calif.	x		x	x	x	x	x	x	x		x	x
Colo. (1)				x				x	x	x	x	x
Conn.	x		x	x	x			x		x		
Del.				x	x	x	x	x			x	x
Fla.	x			x	x		x	x	x		x	x
Ga.			x	x	x			x	x			x
Hawaii										x	x	
Idaho	x	x			x			x	x	x		x
Ill.	x			x	x			x	x		x	x
Ind.						x	x	x	x	x	x	x
Iowa		x		x		x	x	x		x	x	x
Kans.	x	x	x	x	x				x	x	x	x
Ky.			x	x	x				x		x	x
La.				x	x			x	x		x	x
Maine			x	x				x		x	x	x
Mass.	x			x			x	x	x	x	x	x
Md.	x			x	x		x	x	x	x	x	x
Mich.	x		x	x	x	x	x	x	x		x	x
Minn.	x		x	x				x		x	x	x
Miss.				x	x		x	x			x	

Mo.	x		x(2)		x					x	x	x
Mont.	x		x	x	x				x		x	x
N.C.					x		x	x			x	x
N. Dak.	x		x	x	x		x	x		x	x	x
Neb.			x	x	x	x	x	x	x		x	x
Nev.	x	x	x	x	x	x	x	x	x		x	x
N.H.	x			x	x	x	x	x	x		x	x
N.J.	x			x	x	x	x	x	x	x	x	x
N. Mex.	x	x	x	x	x	x				x		x
N.Y.	x			x	x	x	x	x	x		x	x
Ohio			x	x	x			x	x		x	x
Okla.	x			x	x	x	x	x		x	x	x
Oreg.	x		x	x	x	x		x	x		x	x
Pa.	x	x		x	x	x	x	x	x	x	x	x
R.I.	x			x	x		x	x		x	x	x
S.C.				x	x		x	x				
S. Dak.	x		x	x	x		x	x	x			x
Tenn.	x	x		x	x	x	x	x	x	x	x	x
Tex.				x	x		x	x	x	x	x	x
Utah	x		x		x	x		x	x			x
Va.	x			x			x	x	x		x	x
Vt.	x			x		x	x	x	x	x	x	x
Wash.	x			x	x			x	x			x
Wis.		x		x			x	x		x	x	x
W. Va.	x			x	x	x	x	x	x	x	x	x
Wyo.	x			x		x	x		x		x	x

(1). Beekeepers with 25 hives or under are exempt from **all** provisions of the law unless they declare they are selling honey.
(2). Only on requested inspections.

TABLE 3 — U. S. INTERSTATE LAWS

	Entry				Certificate			
State	Prohibited	Permit Required	Must be Quarantined after Entry (1)	Fee	Required with Application	Must Accompany Load	Time Limit for Inspection(1)	Permit in lieu of Certificate
Ala. (2)	x							
Alaska								
Ariz.		x			x		90	
Ark.		x	x		x	x	90	
Calif.		x				x	x	
Colo.(3)		x			x		60	
Conn.						x		
Del.		x			x	x	60	x
Fla.		x			x		30	x
Ga.		x			x		30	
Hawaii (2)		x				x		
Idaho		x		x	x			
Ill.		x			x	x	60	x
Ind.		x				x		
Iowa		x		x	x	x	30	x
Kans. (2)		x		x	x		180	
Ky.		x					30	
La.	x							
Maine		x				x	60	x
Mass.		x				x	60	
Md. (4)						x		x
Mich.	x							
Minn.		x		x	x		60	
Miss.		x			x	x	60	x
Mo.		x		x	x	x	90	
Mont.		x	90	x		x		

N.C.	x			x		60	x
N. Dak.	x		x	x	x	90	x
Neb.	x			x	x	60	x
Nev.	x	x	x			60	
N.H.							
N.J.					x	30	x
N. Mex.					x	90	
N.Y.					x	60	
Ohio	x			x	x		
Okla.					x		
Oreg.		30			x	60	
Pa.	x	x			x	30	
R.I.					x		
S.C.					x	60	
S. Dak.	x		x	x	x	60	
Tenn.	x			x	x	21	
Tex.	x			x	x	60	
Utah	x				x	30	
Va.	x			x	x	60	
Vt.	x				x	ASAP	
Wash.		30 (5)	x (6)		x		
Wis.	x			x		30	
W. Va.	x			x	x		
Wyo.	x		x	x	x	x	

(1) Numbers refer to days.
(2) Special requirements for used equipment.
(3) Beekeepers with 25 hives or under are exempt from **all** provisions of the law unless they declare they are selling honey.
(4) If without accompanying certificate.
(5) If under Quarantine.
(6) Cost of inspection for certificate to be paid by beekeeper.

TABLE 4 — U. S. INTERSTATE LAWS

	Location				Package Bees		Queen Bees	
State	Controlled	Advance Filing Required (1)	Notification Required upon Arrival(1)	Inspection Stamp Required on Hives	Certificate Required (1)	Food Restrictions	Certificate Required (1)	Food Restrictions
Ala.					60		60	x
Alaska								
Ariz.		x	x		x		x	
Ark.	x	x			x		x	
Calif.			3			x		x
Colo. (2)			x					
Conn.					x		x	
Del.					x		x	
Fla.		10			x		x	
Ga.					x		x	
Hawaii					x		x	
Idaho	x	10			x	x	x	x
Ill.			10					
Ind.		x						
Iowa								
Kans.	x	x			x		x	
Ky.					x	x	x	x
La.	x	x			x		x	
Maine					x		x	
Mass.								
Md.								
Mich.					x		x	x
Minn.	x	30						
Miss.					x	x	x	x
Mo.		x			x		x	

Mont.	x	x	x		x		x	
N.C.		x			60	x	x	x
N. Dak.	x	x	x		90		90	
Neb.	x	60	3		x	x	x	x
Nev.	x	x	x		x	x	x	x
N.H.					x		x	
N.J.			x		x		x	
N. Mex.		x	x		x		x	
N.Y.					x		x	
Ohio			10		x	x	x	x
Okla.					x	x	x	x
Oreg.			5		x		x	
Pa.			x	x	x		x	
R.I.								
S.C. (3)					x			
S. Dak.	x			x	x		x	
Tenn. (3)			x		x		x	x
Tex.		10			x	x	x	x
Utah			5		x		x	
Va.				x	x	x	x	x
Vt.		1			x		x	
Wash.			3		x		x	x
Wis. (3)	x							
W. Va. (4)			60		x		x	x
Wyo.	x	x		x	x	x	x	x

(1) Numbers refer to days.
(2) Beekeepers with 25 hives or under are exempt from **all** provisions of the law unless they declare they are selling honey.
(3) Special requirements for used equipment.
(4) Certificate required for honey.

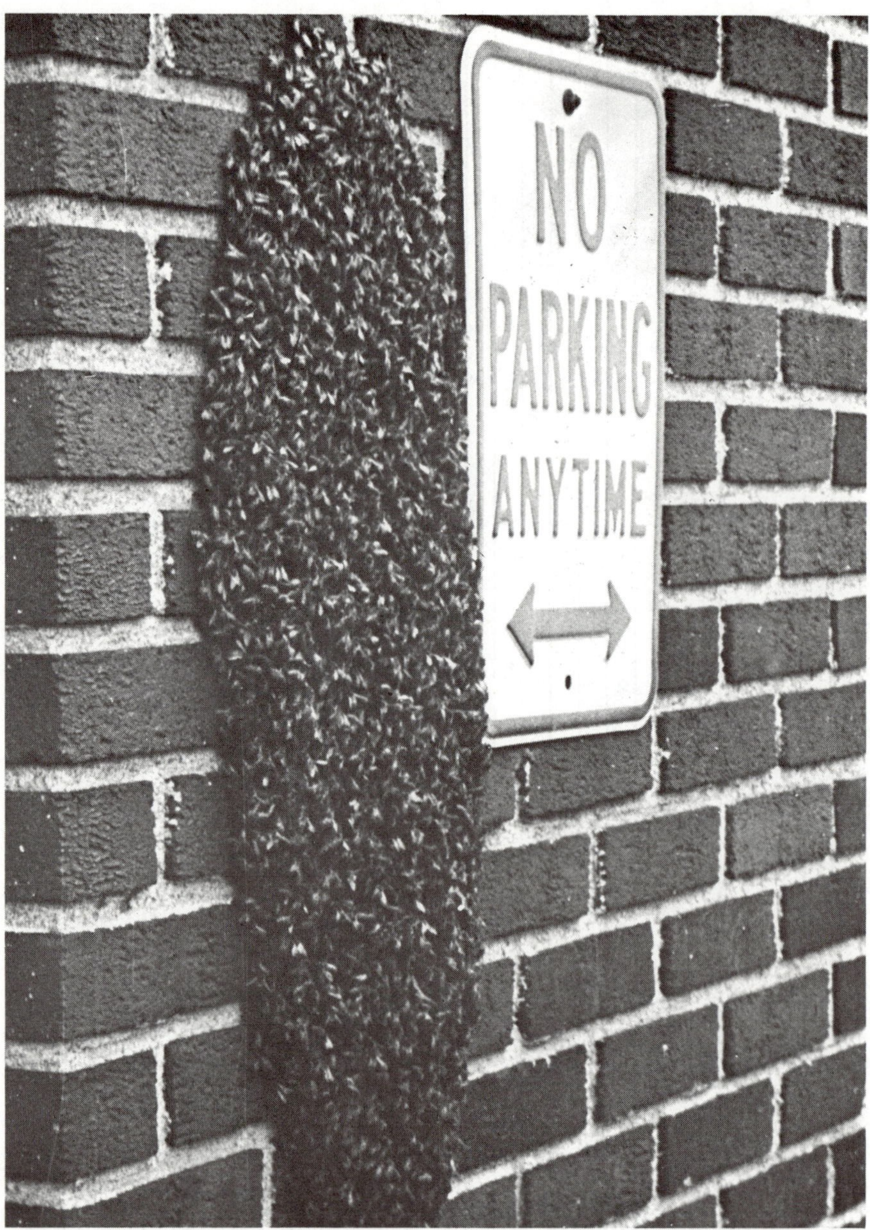

Fig. 2. Prevent swarming. Beekeepers know that swarming is a safe and perfectly natural occurrence but neighbors may be reminded of sensationalized special effects in "scare" movies. Be prepared to handle swarms as quietly and efficiently as possible. Don't try to compete with the movies and other spectacular publicity. Practice an "out-of-sight" — "out-of-mind" style of beekeeping.

CHAPTER II

PROPERTY RIGHTS

You are my honey, honey suckle,
I am the bee
. . . Albert H. Fitz

There is an important distinction between domestic animals and wild animals (ferae naturae) so far as relates to property rights. Over domestic animals — cattle, horses, sheep, barnyard poultry and the like — man may have an absolute dominion and property as over any other useful and valuable chattel.[1] The situation as to property rights or ownership of bees is, however, somewhat varied.

As to the property law of bees, the eminent legal scholar Blackstone, classified them with wild animals; but Blackstone's law was taken from the Greeks and from the Romans, and, curiously enough, there has been practically no change in that law since the days of Plato, and an uninterrupted line of decisions through Greek, Roman, English, French, the Netherlands, and the English common law, down to the decisions in the United States. The probable reason for this set policy being the danger of touching the subject. However, some late rulings have placed the ferae naturae status of bees on the borderline.

Gaius, in the year 160 A.D., wrote in his Commentaries that:

> "In those wild animals whose nature or custom it is to go away and return — as pigeons, bees, and deer, which habitually go into the forests and return — our rule is laid down that only the determination of the intent of returning marks the end of the property in them, and the property in them is acquired by the next one who take (or occupies) them; it is said that when their habit of returning ceases, their intent (or instinct) of returning also ceases."[2]

In the 19th century, Puffendorf expressed his views on property rights of bees in his book, The Law of Nature. He not only stressed the law but included bee instinct and philosophy, thusly:

> "Yet bees are no doubt wild by nature, since their custom of returning to their hive doth not proceed from their familiarity with mankind, but from their own secret instinct; they being in all respects utterly unteachable. It is nevertheless one of Plato's laws — whosoever shall pursue the swarms which belong to others and by striking upon the brass shall draw them with the delightful sound to fix near himself, let him make restitution for the damage. Where he seems to presuppose that the owner of the bees did not follow them when they left his hives, Pliny will have the bees to be neither wild nor tame; others divide them into both kinds. But that, so long as they return to our hives, they are properly our own . . . .[3]

We moderns may wonder at the completeness of property law of bees, and at the attention and time that has been expended, and will expend, upon what we regard as a trivial subject. Nothing is trivial that involves human or property rights. It is to be remembered that, in the days of those whom we term the ancients, the bee occupied a much more important place in the economy of the state than it does now. In Greece, in Egypt, in Judea, and to a somewhat less extent in the Roman provinces, honey was a most important article of commerce.

And in the United States, property in honey occasioned a "war" between two states. This "war," according to Sue Hubbell,[4] was between Missouri and Iowa (not yet a state but part of the Michigan and later Wisconsin Territory) and known as the "Honey War." This bizarre conflict, one of the country's longest, is said to have lasted from 1836 until 1851. It seems a Missourian cut down three bee trees in a disputed area and took the honey from them. He was taken to court in the Iowa Territory, found guilty, and fined $1.50 in costs and damages.

This aroused the ire of Missourians and the "Honey War" had commenced. The issue — whose property was the honey? This was never resolved as indicated by the most notable product of the entire "war" — a poem by John W. Campbell.

The Honey War dispensed with, all other conflicts concerned with property rights to bees and their products have been fought on neutral grounds — the courts. The verdicts have been rendered in numerous states with the main contention being — is the bee ferae naturae or classed as a domestic animal? And the property rights to bees and (wild) honey.

An 1810 ruling in New York discussed qualified property in wild bees:[5]

> "Bees are considered . . . as ferae naturae, but when hived and reclaimed a qualified property may be acquired in them.
> It appears the bees were not hived before they were discovered by Mason, and the only act he did was to mark the tree, The land was not his, nor was it in his possession. Marking the tree did not reclaim the bees, nor vest an exclusive right of property in the finder especially against Gillet. According to the civil law, bees which swarm upon a tree are not private property until actually hived; and he who first enclosed them in a hive becomes their proprietor."

The Supreme Court of Iowa, in 1898, rendered its decision on the property rights in wild bees thusly:[6]

> "Such property rights in wild bees in the woods as there are, are in the owner of the woods; . . . as stated by Blackstone, 'the only ownership in bees is ratione soli (by reason of one's ownership in the soil); and the charter of the forest, which allows every freeman to be entitled to the honey found within his own woods, affords great countenance to this doctrine that a qualified property may be had in bees, in consideration of the property of the soil whereon they are found."

A court in the Empire State, in the year 1916, continued legal rulings on qualified property status of bees, asserting:[7]

> "The qualified property of an owner of a swarm of bees, which flies from the hive, continues so long as he in person or by agent can keep them in sight and possesses the power to pursue them."

The Honorable Judge, in the aforementioned case, further averred that "the case turns upon the question of identity, and of whether the owner, and I shall take the liberty of enlarging the doctrine to include an employee of the owner, kept the bees in sight during the swarm until they alighted; this being really part of the question of identification. I determine that Brown has established the ownership of these bees, and is entitled to recover them, passing without comment the fact that this queen bee occasioned all of us a great deal of trouble by organizing the swarm."

In 1957, the California Superior Court, citing a 1903 Iowa decision, stated that "Bees are ferae naturae . . . . But it has been said that bees, while generally classed as ferae naturae, are so useful and common as to be all but domesticated. However, within the meaning of the municipal code, "bees" are not domestic animals."

The State of Nebraska, as well as other states, treat bees as livestock — for purposes of assessment and taxation. A 1976 Nebraska Supreme Court ruling confirmed that:

> "Bees . . . should be treated the same for assessment and taxation purposes as livestock . . . The Legislature may provide that livestock shall constitute a separate and distinct class of property for purposes of taxation . . . We find no reasonable distinction for tangible personal property tax purposes between livestock and honey bees or other living creatures owned and used for commercial purposes."[8]

This difference in classification of the bee, on one hand ferae naturae but for extraction of revenue a domestic animal, poses a question in the minds of many apiarists. Criminal and civil matters (other than taxing) designate the bee as ferae naturae but for taxing purposes the honey producer is a domestic animal. Again, why?

Nelson E. Bailey, a prominent attorney, in a 1975 article,[9] expounded upon the importance of ownership in criminal activity, particularly larceny of beehives and equipment. The article stressed the value of proving ownership of equipment and the branding of bee supplies and furnishings. In years past, branding acted as a deterrent against cattle rustlers and the same may divert potential "bee rustlers."

Numerous police departments, throughout the United States, have requested home owners to affix brands on their television sets, stereos, and other household equipment likely to be stolen. Even pets, particularly canines, are branded (tattooed) to prove ownership in a case of a 'dognapping.' The branding of frames, hives, smokers, and other beekeeping equipment will be a mark of identification in a larceny matter.

The pilferage of (wild) honey is a crime if it can be shown (proved) the said honey was the property of the owner of the land. An article, written by the author in September, 1977,[10] on the removal of (wild) honey, expresses the viewpoint of a Pennsylvania court:[11]

> "If the bees are his property (landowner), the honey is; for it is the manufacture of his bees. It is the production of what may be called his flock.

> But this requires proof, in the first instance, that the bees are his property. The treeing on his land may form some presumption of it, ...
> It cannot necessarily be inferred therefore, that the honey made in a tree, on another's land, is made by the bees of the owner of the land. Here however, a difficulty occurs; it may not have been the herbage of the owner of the land (herbagium terrae), from whence the liquid was extracted that was made into honey. Bees do not confine themselves to the fruits or flowers growing near, but move to a distance.
> It cannot therefore be made out to a certainty that the honey has been made out of the grass, or vegetables, of the owner of the land on which the tree is. Such occupancy does not give a property.
> A trespass would be committed in coming upon the land . . . .

The above viewpoint was confirmed, in a (wild) honey larceny case, by a New York court.[12]

An 1804 New Hampshire case was somewhat more forceful as to ownership of honey than the case mentioned above:

> "Similarly, a landowner is the owner ratione soli of honey made by bees in a tree on his land, and no property therein may be claimed by another by reason of seeing the bees hive in a tree or finding the honey . . .
> It is much more consonant to our ideas of property to say that the bees and honey in the owner's trees belong to him in the same manner as all mines and minerals belong to the owner of the soil."

CITATIONS

1) *Helsel v. Fletcher,* 225 P.514, 98 Okla.285
2) *Brown v. Eckes,* 160 N.Y.S.489
3) *Ibid.*
4) *Sue Hubbell,* The Venomous Bee, American Bee Journal, February 1978
5) *Gillet v. Mason,* 7 Johns (N.Y.) 16
6) *State v. Repp,* 73 N.W.829, 104 Iowa 305
7) *Id,* 160 N.Y.S.489
8) *Knoefler Honey Farms v. County of Sherman,* 243 N.W.2d.760, 196 Neb.435
9) *Nelson E. Bailey,* The Beekeeper's Role in Criminal Prosecutions of Beehive Thefts, American Bee Journal, November 1975
10) *Murray Loring,* You be the Judge, Gleanings in Bee Culture, September 1977
11) *Wallis v. Mease,* 3 Bin. Pa.546
12) *People v. Hutchinson et als.,* 9 N.Y.S.2d.656, 169 Misc. 465
13) *Fisher v. Stewart,* Smith (N.H.) 60

CHAPTER III
PROPENSITY

Thus may we gather honey from the weed
And make a moral of the devil himself
. . . Shakespeare

The dictionary defines this term as "a natural inclination or bent, a natural tendency." In the field of commerce, it is the duty of the manufacturer to warn a potential customer of the dangerous propensity of its product, if the product has a dangerous propensity. If there is a breach of this duty, the manufacturer may be liable if an injury results from the use of this product. What does this have to do with bees or bee raising?

Merely this — an animal's (bee's) propensity is its natural tendency to behave in a certain manner — more exactly, in a manner likely to endanger the safety of other persons or their property, including other animals. It is not enough that there be potential danger but there must be propensity, that is, a natural inclination to be dangerous.[1]

Frequent reference is made to a vicious propensity which might suggest that propensity inevitably involves a malice or meanness in the animal, an intent to do damage. An Iowa court defined vicious propensity as "a propensity as exists in an animal which might attack or injure the safety of persons without being provoked to do so."[2]

Under the law everyone harboring an animal ferae naturae is charged with knowledge of its habit(s) and evil propensity(ies). However, an Iowa court did not follow the strict interpretation of animal ferae naturae; it looked upon the bee as more of a domestic animal when it ruled:—[3]

> "Those keeping animals of whose mischievous nature everyone is presumed to have knowledge, would exercise reasonable care for the protection of others from their depredations. True, bees may not be confined like the wild beasts. To roam seems to be necessary to their existence. They fly great distances, and if interfered with, or their course obstructed, are likely to resent by the use of their only available weapon. Everyone harboring animal ferae naturae is charged with knowledge of their habits and propensities. But bees, while generally classed as ferae naturae, are so useful and common as to be all but domesticated. Keepers of the apiary have carefully studied their habits and instincts, and control them as certainly as domestic animals."

In Delaware, a similar reasoning was the ruling:—[4]

> "That in this case actual notice must be brought home to the bee owner of the propensity of his animals to do mischief, and there is no proof to that effect."

And in Mississippi the court further expanded upon this principle of classifying the bee as a domestic animal on the issue of propensity:—[5]

> "There seems to be no direct case in our own reports discussing the liability of the keeper of bees for injury done by them but it seems to be the general rule in other states that, as bees are useful to society, and are property of

> value, the ordinary rule as to wild animals, imposing absolute liability for the injuries inflicted by them, is not applicable to bees but the rules of domestic animals; that is, that the owner must know of their vicious tendencies and that the owner is under a reasonable duty to place bees so that they will not come in contact with persons traveling roads and similar places."

As early as 1850, until the present date, an owner of bees who has notice of their vicious character in attacking persons or animals is liable for the ensuing damages. Courts in New York,[6] Florida,[7] Mississippi,[8] and other states have rendered decisions following this legal view. Conversely, those beekeepers who are not put on notice of the dangerous propensity of their bees or, as a reasonable man should have known of their dangerous propensity, may not be liable for damages caused by them. A Delaware court stressed this point of propensity knowledge in its decision:—[9]

> "It was further pointed out that there was no proof that the bees were kept in an improper place or in an improper manner, or that the beekeeper had any notice of the propensity of the bees to do mischief."

It is a known fact that courts render binding decisions in settlement of controversies but infrequently offer advice. So it was an enlightenment to read of legal assistance to animal (bee) owners on the issue of propensity.[10]

> "The owner or keeper of a domestic animal is bound to take notice of the general propensities of the class to which it belongs, and also of any particular propensities peculiar to the animal itself or which he has knowledge or is put on notice; and in so far as such propensities are of a nature likely to cause injury he must exercise reasonable care to guard against them and to prevent injuries which are reasonably to be anticipated from them . . . ."

Whereas propensity may be unfamiliar to some attorneys, particularly involving bees, a practice pointer may be helpful. In those proceedings where liability of a beekeeper may depend upon some notice on his part of the vicious propensities of the particular bees that were alleged to have caused the injuries or damage, counsel for the beekeeper may find it useful, where warranted by the evidence, to focus upon the circumstance that there had been no other incident of injury or damage by the bees despite the fact that they had been maintained at the same location for several years.[11]

CITATIONS

1) — *Talley v. Travelers Ins. Co.*, La.App., 197 So. 2d. 92
2) — *Mellicker v. Sedlacek,* 179 N.W.197, 189 Iowa 946
3) — *Parsons v. Manser,* Iowa App. 93 N.W. 86
4) — *Petey Mfg. Co. v. Dryden,* Del. App., 62 A.1056
5) — *Ammons v. Kellogg,* 102 So. 562, 137 Miss. 551
6) — *Earl v. Van Alstine,* 8 Barb 630
7) — *Ferreira v. D'Asaro,* Fla. App., 152 So. 2d. 736
8) — *Id,* 102 So. 562, 137 Miss. 551
9) — *Id,* Del. App., 62 A. 1056
10) — *Loftin v. McCrainie,* 47 So. 2d. 298
11) — 86 A.L.R. 3d. 832

CHAPTER IV

NUISANCE

All the breath and the bloom of the year is the bag of one bee.

. . . . Elizabeth Barrett Browning

One legal authority, name unknown, speaking on this subject, recognized: —

> "There is perhaps no more impenetrable jungle in the entire law than that which surrounds the word 'nuisance'. It has meant all things to all men, and has been applied indiscriminately to everything from an alarming advertisement to a cockroach baked in a pie."

Our courts have been wrestling with this word for years, for centuries, and it is little wonder. Nuisance is one of those pesky words with only a handful of basic definitions — obnoxious, annoying, offensive, disturbing — but with a fistful of possible interpretations, nuances, explanations, shades of meaning, et cetera, et cetera, et cetera. Like obscenity, nuisance is a phantom concept. Its substance differs from person to person, from place to place, from animal to animal.

Who among us has not been roused from restful sleep by the untimely howling of a neighbor's dog? Of course, if you happen to be the neighbor and a sound sleeper, then the howling would not be a nuisance, would it? It all depends.

Still, the law, in its majestic striving to keep the peace, has made some progress in defining nuisance.

As a general rule, animals (bees included) are not regarded as nuisances as such, but they, or the places where they are kept or housed, may be or become nuisances under the circumstances of the particular case.[1] The keeping of a large number of animals in a residential area in a city or near a public thoroughfare may be a nuisance, where the numbers may be detrimental to the comfort of those living in the neighborhood or passing by.[2]

It is likewise a general rule, as written by a prominent attorney,[3] "that an owner or lawful occupant of property is legally free to use his property as he sees fit, regardless of objections from a neighbor, so long as the use does not violate any specific ordinances or statutes. There is, however, a limitation to this general rule. An occupant or owner is not permitted to use his property in a manner that seriously interferes with a neighbor's legal right to use his own property as he sees fit. Where an owner uses his property so as to seriously interfere with a neighbor's right to use his property, the owner who is causing the interference is maintaining what the law calls a 'nuisance'."

Keeping bees in a manner to interfere seriously with the rights of a neighbor or adjoining landowner may constitute a nuisance.

The courts, in various areas, have been more direct in their decisions, as in Canada where: —[4]

> "The right of a person to enjoy and deal with his own property as he chooses is controlled by his duty to so use it as not to effect injuriously the rights of others, and in this case it is a pure question of fact whether the bee raiser collected on his land such an unreasonably large number of bees, or placed them in such position theron, as to interfere with the reasonable enjoyment of the landowner's property."

In New York, the court gave its ruling but gave the apiarist an alternative: —[5]

> "That defendants had maintained, and were still maintaining a large number of hives of bees, kept in an open lot immediately adjoining plaintiff's dwelling house, and that at certain seasons they were a source of constant annoyance and discomfort to plaintiff and his family, greatly impairing the comfortable enjoyment of the property; also that the bees could be removed without material injury, to a locality where neighbors would not be disturbed by them."

Florida follows the trend of the general rule, forewarning bee raisers, as: —[6]

> "Keeping of bees in a manner to interfere seriously with rights of adjoining land owners may constitute a nuisance."

An early Arkansas ruling was on generalities — the case had an 1889 dating: —[7]

> "Neither keeping, owning, nor raising bees in a city is itself a nuisance. . . . that bees could become a nuisance in a city, but whether or not they were a nuisance would be dependent on the facts in each case; they would not automatically be a nuisance just because they are bees."

It is obvious that the keeping of bees is not unlawful in and of itself. Rather, for some liability to attach to keeping bees, it must be established that the beekeeper violated a standard of due care. The test which seems to apply is that of reasonableness.[8] This test was applied in a Pennsylvania proceeding: —[9]

> "although the keeping of bees is not a per se nuisance, an unreasonable number of bees in an unreasonable place may constitute a nuisance."

Expanding upon the aforementioned theory, a court in New York held: —[10]

> "that repeated entry of bees onto the land of another resulting in a continuous and material annoyance rendering such persons' enjoyment of their property uncomfortable constituted a nuisance subject to restraint in equity."

A 1946 Pennsylvania case followed its predecessors in ruling that: —[11]

> "the keeping of an unreasonable number of bees in an unreasonable place might contribute a nuisance simply because of the manner in which the maintenance of bees was conducted . . . . whether the complainants were subjected to unreasonable inconvenience and annoyance as a result of the owner's beekeeping."

In a case of recent vintage, 1979, the Ohio court interspersed legality with understanding on the question of a nuisance: —[12]

> "Defendants may maintain two hives within fifty feet of the extracting building contemperaneously with extracting honey; and, defendants may maintain hives from packages of bees for the purpose of starting new hives for a total period not to exceed twenty-one days during the months of April or May in each year. Defendants shall give plaintiffs notice of at least fourteen days before such hives are used.
> That defendants not maintain hives or colonies of bees within one-half mile of plaintiffs residence save and except as stated."

The majority of cases summarized heretofore have been a courtroom confrontation betweeen the apiarist and an adjoining neighbor. The issue — was the beekeeper responsible for the 'private nuisance'. In other cases the term 'public nuisance' has been used. Is there a difference? In legal vernacular there is a distinction. Court cases point this out, as: —

> "A *private nuisance* — a nuisance which violates only private rights and produces damages to one or to no more than a few persons."[13]

> "A *public nuisance*— a condition of things which is prejudicial to the health, comfort, saftey, property, . . . of the citizens at large resulting either from an act not warranted by law, or from neglect of a duty imposed by law."[14]

A New York court, in a 1954 decision, enlarged on a public nuisance: —[15]

> "City resident had right to maintain colony of honey bees upon his premises but was charged with duty of maintaining them in such manner that they would not annoy, injure or endanger comfort, repose, health or safety of any considerable number of persons insecure in use of their property, and if he omitted to do so he was subject to indictment under statute of crime for maintaining a public nuisance."

The basic difference between private and public nuisance can be simply stated: Private nuisance — individual or few persons affected; public nuisance — a considerable number of persons involved.

Oftimes the law, in its majestic striving to seek equity, has favored beekeeping as an economic and/or socially desirable occupation.[16] It has been stated that bees are among those usually harmless 'domestic' animals which cannot be constantly confined without interfering with their economic usefulness, and that, accordingly, in certain locations they have been allowed as a matter of tradition to run at large without being viewed in the same light as livestock . . . . .[17]

The law looks with more favor upon the keeping of animals that are useful to society than those that are useless. This was exemplified in Pennsylvania where: —[18]

> "no one is entitled to absolute quiet in the enjoyment of his property, and that an individual can only insist upon a degree of quiet consistent with the standards prevailing in the location in which he lives."

In deciding whether or not bees are a nuisance and will be ordered moved away, a court will consider the beekeeper's business reason for having the bees where they are. If the bees are a nuisance and the beekeeper has another location a short distance away where the bees can be safley placed, the court almost certainly will order them moved. But, if the beekeeper has no other place to hive his bees in that area, and the beekeeper's business substantially depends on his acquiring the honey flow where located, and the nuisance to a neighbor is not extremely unbearable, there may be a good chance the bees will be permitted to remain.

Generally, the more compelling the business reasons are for keeping bees where they are, the more inconveniences an adjoining landowner will be compelled to tolerate before a court will rule the bees a nuisance and order them to be removed.[19]

An interposing legality, lately assuming importance, is the phrase "coming to the nuisance." In years past, the doctrine was "that one who moves into the neighborhood of an existing nuisance cannot complain of it." Today the weight of authority is against that doctrine.

The mushrooming of urban developments into rural areas and where more and more homes, trailer parks and commercial ventures are spreading into the country where bee raisers thought themselves secure, "coming to the nuisance" is an ever-increasing problem. A Wisconsin court, speaking of the horse, but just as easily could be speaking of the bee, rendered its ruling on this difficult question: —[20]

> "We cannot look with favor on one who knowingly commenced a restaurant business close to an area that was being used to stable over 50 horses. It is understood that one is not barred from relief in the courts merely because of 'coming to the nuisance', but it is a factor which bears upon the question of whether the restaurant owner used his land reasonably under the circumstances. We must conclude that his determination to use the land for restaurant purposes — at least of the drive-in type that he built — was an unreasonable use under the circumstances . . . . .
> This was predominantly rural area, and what would be a nuisance on the Capitol Square would not be a nuisance in the country. A nuisance may be merely a right thing in the wrong place, like a pig in the parlor instead of the barnyard."

Discussion has centered on the apiarist, as it should. What are the rights of the landowner situated adjacent or close by a colony of bees? Generally speaking, one claiming ownership or otherwise lawful possession of property, then one has the right not only to the unimpaired condition of the property itself, but also to some reasonable comfort and convenience in its occupation. Enjoyment of property is inseparable from ownership of the property.

The question of bee stings is properly covered in the chapter on Negligence. A Florida court, in its wisdom, justified bee injuries into the realm of nuisance: —[21]

> "Bees were a pastime hobby with this apiarist. He had seven hives which he kept for the "honey" and for his enjoyment. One particular day a swarm of bees buzzed over to the neighbor's yard and commenced bothering the dog. Seeing her animal annoyed, its owner went to the dog's rescue and was

severely stung by a large number of these insects. Medical attention was immediately necessary as shock was imminent.

The dog's owner sued the beekeeper for monetary damages, as the result of injuries received from multiple bee stings. Her husband joined in the action and sought derivative damages, not only for his wife's medical expenses, but for loss of her services and consortium.

The plaintiff's evidence pointed out that the beekeeper was negligent keeping bees in close proximity, 31 feet, to the neighbor's residence; by failure to properly house the bees, having supposed knowledge of their propensity to inflict injury; by failure to control or contain them when they become angered and aroused to violence; and that such keeping and handling of the bees constituted a nuisance.

"I am sorry," answered the beekeeper, "for the stings received by my neighbor, but she was responsible for her own injuries. She had been living with my bees for quite a time and had never received any previous stings. She brought peril upon herself by swatting at them. If she had let them alone, they would have returned to their hives."

"I agree with you," decisioned the Court, "your bees had no propensity to cause injury but your keeping of them constituted a nuisance."

The liability of an owner for injury inflicted by bees is for negligence in the location and manner of keeping them. Anything which annoys or disturbs one in the free use, possession or enjoyment of his or her property or which renders its ordinary use or occupation physically uncomfortable may become a nuisance.

And damages may be awarded for loss or injury proximately resulting from maintenance of a nuisance."

A learned New Jersey judge brought out the fact that roosters will strut, wax indignant and crow. Yet they, too, have moments of grave serenity. Hens will perambulate, scratch and cackle. Nevertheless the keeping of chickens is not, in all circumstances, a nuisance as such.

In a similar vein, bees will buzz, roam and, when disturbed, occasionally sting. Yet they, too, have lasting moments of hive serenity. The keeping of them is not, in most circumstances, a nuisance as such.

CITATIONS

1) — *Smith v. Costello,* 290 P.2d.742, 77 Idaho 205
2) — *Boudinat v. Smith,* Okla.Cr., 340 P.2d.268
3) — *Nelson E. Bailey,* American Bee Journal, December 1975
4) — *Lucas v. Pettit,* 12 Ont.L.Rep.448
5) — *Olmstead v. Rich,* 6 N.Y.S.826
6) — *Ferreira V. D'Asaro,* Fla.App., 152 So.2d.736
7) — *City of Arkadelphia v. Clark,* 11 S.W.957, 52 Ark.23
8) — *Riffle et al v. Moore et al,* Case 22836, Union County, Ohio
9) — *Allman v. Rexer,* 21 Pa.D&C 431, 82 Pittsb.Leg.J.367
10) — *Id,* 6 N.Y.S.826
11) — *Holden v. Lewis,* 33 Del.Co.458, 56 Pa.D&C 639
12) — *Id,* Case 22836 Union County, Ohio
13) — *Riggins v. District Court,* 51 P.2d.645, 89 Utah 83

14) — *Nicholas v. Commonwealth,* 226 S.W.2d.796, 312 Ky.171
15) — *People v. McOmber,* 133 N.Y.S.2d.407, 206 Misc.465
16) — *Earl v. Van Alstine,* 8 Barb 630
17) — *Restatement, 2d., Torts* Sect.518, comment
18) — *Id,* 21Pa.D&C 431, 82 Pittsb.Leg.J.367
19) — *Id,* American Bee Journal, December 1975
20) — *Abdella v. Smith,* 149 N.W.2d.537, 34 Wis. 393
21) — *Id,* Fla.App., 152 So.2d.736

Fig. 3. Bees fill their honey stomach prior to swarming and may easily be handled by a competent beekeeper. Your calm attitude toward bees will also calm anxious neighbors who may be watching.

CHAPTER V

NEGLIGENCE

Fruitless as the celebrated bee who wanted to swarm alone
. . . G. K. Chesterton

Were one to describe negligence in one word, it would be carelessness. A more complete legal explanation would be: "The omission to do something which a reasonable man, guided by those considerations which originally regulate human affairs, would do, or doing something which a prudent and reasonable man would not do."[1] It is also defined as "the failure to exercise a certain degree of care, as required by law, for the protection of other persons or property, including animals."

One has the right to own and keep bees. However, there is also the obligation to see that the bees do not endanger other people, engaged in ordinary and lawful activities, or endanger their property. If the bees do injure person or property, one may find himself on the receiving end of a lawsuit. The basic charge — negligence.

Of course, the charge of negligence can come from the bee owner as the result of injury or death to his bees by another person.

Negligence cannot be presumed from the mere happening of an injury. The burden is on the plaintiff who alleges negligence to produce evidence from which the judge or jury can find that the defendant was guilty of negligence which was a proximate cause of the injury. The evidence produced must prove more than a probability of negligence and any inferences therefrom must be based on facts, not on presumptions.

In most cases involving negligence, the issue of liability — one's legal responsibility or obligation — is invariably intertwined. This liability can be the result of injury by bees as well as injury to bees.

Very often, in cases based on negligence, the issues of Propensity, Nuisance, Trespass, and others may be encountered. These isssues are treated more thoroughly in their appropriate chapters.

Courts determining a beekeeper's liability have generally not applied the common-law rule of absolute liability that has been applied to the keepers of wild animals, and the courts in a number of cases have expressly held that beekeepers are liable only if they have been found to have acted negligently,[2] as in Florida:[3]

> "An action growing out of injuries sustained by a lady from multiple stings from apiarian's bees. Liability of bee owner for injury inflicted by bees is for negligence in location and manner of keeping them, and not on basis of common-law liability for keeping wild animals."

Similarly a Mississippi court ruled:[4]

> "the rule of absolute liability that was generally applied to liability for injuries inflicted by wild animals was not applicable, but the rule applicable

to injuries caused by domestic animals would be applied . . . (1) in order to be held liable, the owner of the bees must know of their vicious tendencies, and (2) such owner is under a reasonable duty to locate the bees in a place where they will not be in contact with persons traveling roads and similar places."

As early as 1850, in New York, and 1903, in Delaware, the common-law rule of absolute liability was not applied relative to bees. The New York case said:[5]

"liability does not depend upon the classification of the animal as "ferae naturae". . ."

And in Delaware:[6]

"though bees are ferae naturae, they are kept for the use and convenience of man, and are excepted from the rule of law that keeping of all animals ferae naturae presumes negligence"

In cases involving bee stings, the question of "cause" may become important. Specifically, it may be difficult to determine exactly which bees, if they were bees, caused a particular injury. The beekeeper is to focus upon any uncertainty that exists as to the identity of the offending bees. It is worth noting, however, that even where the identity of the bees causing particular damage or injuries involves some conjecture, this problem may not in itself constitute grounds for the avoidance of liability.[7] This was shown in Mississippi where:[8]

"In an action for damages against the owner of bees for their vicious attack upon animals, evidence that the apiarian had a large number of bees in the immediate vicinity of the attack, with proof that the bees came from the direction in which the hives were located is sufficient for the jury to infer that the apiarian's bees did the damage."

In recent years, the majority of legal proceedings, in which negligence is the paramount issue, involves the spraying or dusting of fruits, vegetables or other farm products from an airplane or other mechanical device. The spraying or dusting operator and/or landowner may become liable for resulting damage if the individual(s) negligently spreads liquid or powder known to contain a poison in such a manner as to endanger the lives of bees, other animals or property of another person in the immediate vicinity.[9]

Before dwelling on negligence cases in which bees and (sprayed) poisonous liquids and dusts came together, the term poisonous is defined by an Arizona court:[10]

"The bee owner alleged that the Dutrox spray or dust was poisonous, and that the poison killed and damaged his bees. The evidence that it was poisonous was very meager aside from the fact that it killed most of the bees that it contacted. If it killed the bees, it was because they inhaled it. It was poisonous to them."

The destruction of bees and honey, via the spraying of poisonous materials, negligently or not, commands a number of case reports. In California:[11]

"It was found that the bees came to their death by means of the poisonous dust floating from the field to and into the hives, and not by their going in search of honey to the blossoms on the field which was being dusted . . . bee owner testified that he found a fine gray dust in the hives and on the dead bees. All the bees were dead, including the nurse bees and the queen bee. There is evidence that when bees go to a field that has been dusted with calcium arsenate it takes at least three weeks to kill the workers of a swarm; that the nurse bees and the queen bee never leave the hive during the honey making season. The longest period that could have elapsed between the dusting operation and the finding of the dead bees was fourteen days. Three weeks had not elapsed between the dusting operations and the finding of the bees, all dead, including the nurse bees and the queen bee. This evidence supports the inference . . . that the bees were killed by dust which floated into the hives and not from dust obtained by the bees from the flowers of the melon field.

Landowner and airplane operator knew, or should have known, that the light dust projected under pressure onto the melons would float in the air. There is evidence that a light breeze was blowing during the dusting operations. They should have known that the dust would float for a considerable distance when propelled by such a breeze. Dusting material containing a poison that would kill bees was used. Under the conditions prevailing at the time they should have forseen the ensuing damage to the bee owner."

A later case, in another western state, followed in a like vain as the decision in California:[12]

> "The testimony as a whole shows . . . that owner's bees came to their death as a result of the dusting with poisonous arsenic compound of the adjacent cotton crop by the dust falling and settling around and upon the apiary and premises of bee owner due to the negligent acts of the landowner and airplane operator."

The Supreme Court of Iowa rendered a verdict against an Iowa city for negligent spraying with a poisonous substance. The high court said that:[13]

> "Where city rented airbase property to apiarian, it was engaged in its proprietary capacity, and as such city had liability arising from landlord and tenent relationship, including liability for negligent grasshopper spraying operation which killed apiarian's bees, and spoiled his honey and hives located at municipal airbase on an acreage rented from city."

Conversely, the New York State Conservation Department was absolved of negligence in its spraying of DDT in oil by the state court. They averred:[14]

> "Proceedings against State for damages for loss of bees allegedly destroyed by aerial spraying of DDT in oil by the State Conservation Department . . . where the facts proven show that there are several possible causes of an injury, for one or more of which the State was not responsible, and it is just as reasonable and probable that the injury was a result of one cause or the other, apiarians cannot have a recovery since they have failed to prove that the negligence of the State caused the injury."

Finally, in this category, a California court ruled "insufficiency" to a charge of negligence:[15]

> "In apiarian's complaint nowhere is it alleged that the dust settled upon hives or upon the property of bee owner, but only that the bees came in contact with the dust while on landowner's property and property of others . . . this is not a sufficient basis upon which to predicate liability on the part of pepper grower and airplane operator."

Transportation carriers, be it rail, motor, or air, moving bees from place to place, are required to carry their cargo safely and will be held liable for injury to the bees if due to their negligence. Most court decisions respond in damages to the injured beekeeper, as in Alabama:[16]

> "In bee owner's action for damages resulting from death of bees in transit via interstate carrier . . . to make out a prima facie case under . . . a bill of lading of a common carrier . . . bee owner has the burden to prove the material facts alleged, that is, the receipt of the goods (bees) by the carrier for transportation to the point of destination, for a reward, and that the goods were injured in transit with resultant damages. Upon such proof a prima facie case. including negligence, is established against the carrier."

Although prior in time by many years, a Texas case had a ruling as that in Alabama:[17]

> "Where a carrier's agent was informed that bee owner desired the use of a car for the shipment of bees, but furnished a car which was unsuitable therefore, and so defective that the bees were injured after the car had left the carrier's road and were in the possession of a connecting carrier, the carrier was liable therefore."

Some rulings in foreign countries, relating to a beekeeper's liability for injuries inflicted by his bees, have followed the decisions as that reached in many state courts of the United States. In Canada:[18]

> "If the number of bees was unreasonable, or if they were so placed as to interfere with his neighbor or in the fair enjoyment of his rights, then what would otherwise have been lawful becomes an unlawful act . . . the bees, because of their number and situation, were dangerous to plaintiff. The defendant was acting unlawfully, and he is liable for the injury flowing directly from such unlawful act."

And in Ireland:[19]

> "It appeared that the bee owner kept large numbers of bee hives close to the boundary fence between his and his neighbor's yards; that on the occasion in question the bee owner smoked his bees out to obtain their honey; that the bees, irritated by the smoke, swarmed on the neighbor and his horse in a near-by field, and that the horse, stung by the bees, dragged the neighbor and threw him violently against a wall, causing him severe injuries. A verdict against the bee owner was sustained."

A nervous and tense situation may be experienced by the beekeeper when he is asked to testify in court — his bees were injured by a neighbor's spraying operation. On the witness stand he will be asked questions to establish facts and circumstances that his neighbor failed to use reasonable care in spraying crops near his bee hives. A list of questions and answers are incorporated to point out a courtroom scenario that may occur:

Q. What is your occupation?
A. I am a beekeeper.
Q. Where do you keep your bees?
A. Most of them are near a tamarack grove on my land. But I also kept one hive near the eastern line of my property, across a dirt road from Mr. Smith's tomato field.
Q. What is Mr. Smith's occupation?
A. He is a farmer who raises tomatoes, cantaloupes, and other crops.
Q. Did Mr. Smith ever speak to you about your keeping bees near his tomato crops?

A. Yes, sir. It was at his request that I set a hive there.
Q. Did Mr. Smith give you a reason to keep your bees near his tomato crops?
A. Yes. So that they could fertilize the plants. Bees are the most efficient means of pollination.
Q. Who was responsible for the maintenance of the hive?
A. I was.
Q. Did you have any conversation with Mr. Smith about the safety of your bees?
A. Yes, sir. When I placed the hive near his property, I asked him to warn me if he ever sprayed his field with insecticide.
Q. What, if anything, did he say when you asked him to notify you about any possible sprayings?
A. He said he would, that he knew a bug killer would also injure bees.
Q. Did any sprayings of Mr. Smith's land ever take place after this converstaion?
A. Yes. An airplane dusted Mr. Smith's tomato field on the third day of June.
Q. Did Mr. Smith tell you of this spraying?
A. No, sir. He did not.
Q. Did Mr. Smith offer any reason for his not telling you?
A. He apologized later, but said that it-had just slipped his mind.
Q. What were the weather conditions the day of the spraying?
A. There was a continuing strong wind — about twenty miles an hour coming from the east, that is, from Mr. Smith's land to my own.
Q. What was the condition of the hive before the spraying?
A. The hive was clean, and all the bees were healthy.
Q. What was the condition of the hive after the spraying?
A. When I inspected it two hours after the spraying, the hive was covered with a grayish-white dust which I recognized as a chemical spray.
Q. What was the condition of the bees after the spraying?
A. They were all covered with the grayish-white dust, including the queen and the nurse bees, which never leave the hive during the honey-making season. Some of the bees were already dead when I inspected the hive, others were weak and apparently sick.
Q. Had you been told of the spraying, were there any means which you could have taken to avoid the danger?
A. Yes, sir. I would have moved the hive as quickly as I could have placed it on my truck.[20]

(The aforementioned questions and answers have been compiled from evidence taken from several cases. The name Mr. Smith is fictitious.)

CITATIONS

1) — *Hulley v. Moosbrugger,* 95 A.1007, 88 N.J.161
2) — *American Law Reports,* 86 ALR 3d.829
3) — *Ferreire v. D'Asaro,* Fla.App., 152 So.2d.736
4) — *Ammons v. Kellogg,* 102 So 562, 137 Miss. 551
5) — *Earl v. Van Alstine,* 8 Barb 630

6) — *Petey Mfg. Co. v. Dryden,* Del.App., 62 A.1056
7) — *Id,* 86 ALR 3d.829
8) — *Id,* 102 So. 562, 137 Miss.551
9) — *Lenk v. Spezia,* 213 P.2d.47, 95 C.A.2d.296
10) — *S. A. Gerrard Co. Inc. v. Fricker,* Ariz. App., 27 P.2d.678
11) — *Miles v. A. Arena Co.,* Cal.App., 73 P.2d.1260
12) — *Lundberg v. Bolon,* 194 P.2d. 454, 67 Ariz.259
13) — *Brown v. Sioux City,* Iowa App., 49 N.W. 2d.853
14) — *Newton v. State,* 222 N.Y.S.2d.959, 31 Misc.2d.48
15) — *Jeanes v. Holtz,* Cal.App., 211 P.2d.925
16) — *Anderson v. Railway Exp. Agency,* Ala.App., 39 So.2d.689
17) — *International & G. N. R. Co. v. Aten,* Tex. App., 81 S.W.346
18) — *Lucas v. Pettit,* 12 Ont.L.Rep.448
19) — *O'Gorman v. O'Gorman,* 2 Ir.R.573, K.B.D.
20) — *Proof of Facts,* The Lawyers Co-Operative Publishing Co.

Fig. 4. This beekeeper is wearing protective clothing and has used a smoker to calm the bees during a routine examination of the hive.

CHAPTER VI

CONTRIBUTORY NEGLIGENCE

Love, in my bosom, like a bee
Doth suck his sweet
. . . . . Thomas Lodge

This issue is a defense, usually brought forth in negligence actions, by the defendant. The cases which have been decided on contributory negligence have been, in the majority, decided against the interest of the beekeeper.

In a legal sense, contributory negligence is "a breach of duty on the part of the plaintiff in an action to recover damages for negligence to exercise the standard of care, which is ordinary care, the care that a reasonably prudent person would exercise, for his own saftey, such breach constituting a defense, in the absence of legislation to the contrary, where it was a legally contributing cause of the injury."[1]

As noted in the chapter on Negligence, negligence is doing something which a prudent and reasonable man would not do. On the other hand, contributory negligence is doing something or omitting to do something, on the part of the complaining party, in concurrence with animal owner's negligence, which through want of ordinary care, is the proximate cause of an injury.[2]

Thus, if the owner of bees had full knowledge of anticipated or actual performance of acts which will result in injury, and he fails and refuses to exercise reasonable precaution to protect them, he is guilty of contributory negligence which will preclude him from receiving damages.[3] In some jurisdictions a partial recovery of damages may be awarded.

The reasoning behind this legal principal is that if the one whom a suit is brought against is expected to exercise a reasonable amount of care to avoid harm to others, or their property, then the complaining party should be expected to exercise a reasonable amount of care to avoid harm to himself or to his property.

An example will explain:

> "John Jones owns a dog, which bites Sam Smith, who then sues John Jones. Sam proves that John knew his dog was vicious, yet negligently let the dog run loose. However, John proves that Sam also knew the dog was vicious, yet negligently went near enough to the dog to be bitten. Sorry Sam, but that happens to be contributory negligence — you lose."

A California proceeding will state in detail the issue of contributory negligence mentioned heretobefore:[4]

> "The plaintiff maintained two apiaries not too far from each other. He was an experienced beekeeper, having been engaged in the business for some thirty-eight years. In July and August dusting was preformed on adjacent tomato farms by blowing an insecticide upon the vines and crops from an airplane.

As a probable result of a poisonous insecticide being used, the beekeeper found many dead bees in his yard and about the hives. He stated that a powder was found upon the hives and bees, and which was analyzed and found to be poisonous. And further, the beekeeper testified the airplane operators released the poisonous insecticide powder from their craft when the wind was blowing directly toward the place where the hives were stationed. And the poisonous compound drifted to and upon his hives and bees.

The allegations, as presented by the beekeeper, were emphatically denied by the airplane operators.

They testified that before dusting the various tomato crops they personally notified the beekeeper of their intention to do so. Offers were made to use their trucks to assist the beekeeper in removing his hives to a safe place, but that he refused to take any means of protecting his bees. Saying with respect to a particular crop, that, "If we dusted that job he would sue us." They further stated that just before dusting the tomato crop adjacent to the beekeeper's home place, with knowledge of their intention to do so, the beekeeper removed the screens which he had preciously placed over the hives. And further, the only insecticide which they used was pink in color.

After all the evidence had been presented by both parties, the justices gave their decision.

The foregoing evidence not only fails to satisfactorily show that it was the insecticide which was used by the airplane operators to spray the crops of tomatoes that killed the beekeeper's bees, but there is substantial proof that, with full knowledge on the part of the beekeeper that it was to be used, he failed and neglected to exercise ordinary precaution to either remove his hives or to adequately screen them during the periods of dusting, for protection of the bees. He was fully aware that farmers in that vicinity produced tomatoes, which they were accustomed to spray or dust in the summer months of each year to protect the crops from worms, insects and other detrimental pests. The beekeeper knew that fact. We are of the opinion that the beekeeper lost his bees as a result of his own contributory negligence.

When an owner of bees contributes proximately to their injury or death, he is thereby barred from recovery . . . . .

We are constrained to hold there is ample evidence in this case to support the findings that the plaintiff is barred from recovery by his contributory negligence."

Although no single test is applicable in determining whether one entering upon premises where bees are present, and is injured, is guilty of contributory negligence, the important factors in determining that question are whether the person has a justifiable right to enter the premises and the manner in which the entry is made. It has been held generally that one is not guilty of contributory negligence merely because he is a trespasser on the premises where the bees are present.[5]

The question whether one is guilty of intentionally provoking bees and is thus barred from recovering for a resulting injury by the bees is generally considered a question of fact. But it has been brought out that the injured person intended to do the very thing which brought on the injury.[6] Also, it has been held that the act provoking the bees must be committed at or near the time of the bees' attack.[7]

In this discussion of contributory negligence it would be remiss if mention is not made of the doctrine of assumption of risk. The Latin maxim "volenti non fit injuria" which translated says "That to which a person assents is not regarded in law as an 'injury'." In modern terminology, assumption of risk is defined as: "one, who voluntarily exposes himself to an obvious, known, and appreciated danger, assumes the risk of injury that may result therefrom and may be precluded from recovery of damages."

One with some knowledge and appreciation of the propensities of bees, who is warned of the same, and is injured by mingling with them, may, in a negligence proceeding, not be awarded the damages sued for — contributory negligent.

To establish the defense of assumption of risk in a suit by an employee, injured by bees, against his employer, it has to be established that the risk was one ordinarily incident to the employment. Or, the employee knew of the dangerous propensities of the bees and yet continued in the employment which exposed him to the risk.[8]

A general rule is that an employee who has knowledge, either actual or constructive, of the dangerous or vicious propensities of the bees owned by his employer with which he has to work, for which he has to care, or with which he otherwise comes in contact, assumes the risk of injury by such bees.[9]

CITATIONS

1) — *Brakensiek v. Nickles,* 227 S.W.2d.948, 216 Ark.889
2) — *Honaker v. Crutchfield,* 57 S.W.2d.502, 247 Ky.495
3) — *Lenk v. Spezia,* 213 P.2d.47, 95 C.A.2d.296
4) — *Ibid.*
5) — *Sanders v. O'Callaghan,* 82 N.W.969, 111 Iowa 574
6) — *Fake v. Addicks,* 47 N.W.450, 45 Minn.37
7) — *Id,* 82 N.W.969, 111 Iowa 574
8) — *Russell Creek Coal Co. v. Wells,* 31 S.E.614, 96 Va. 416
9) — *Corley v. Hubbard,* 260 N.W.551, 129 Neb.38

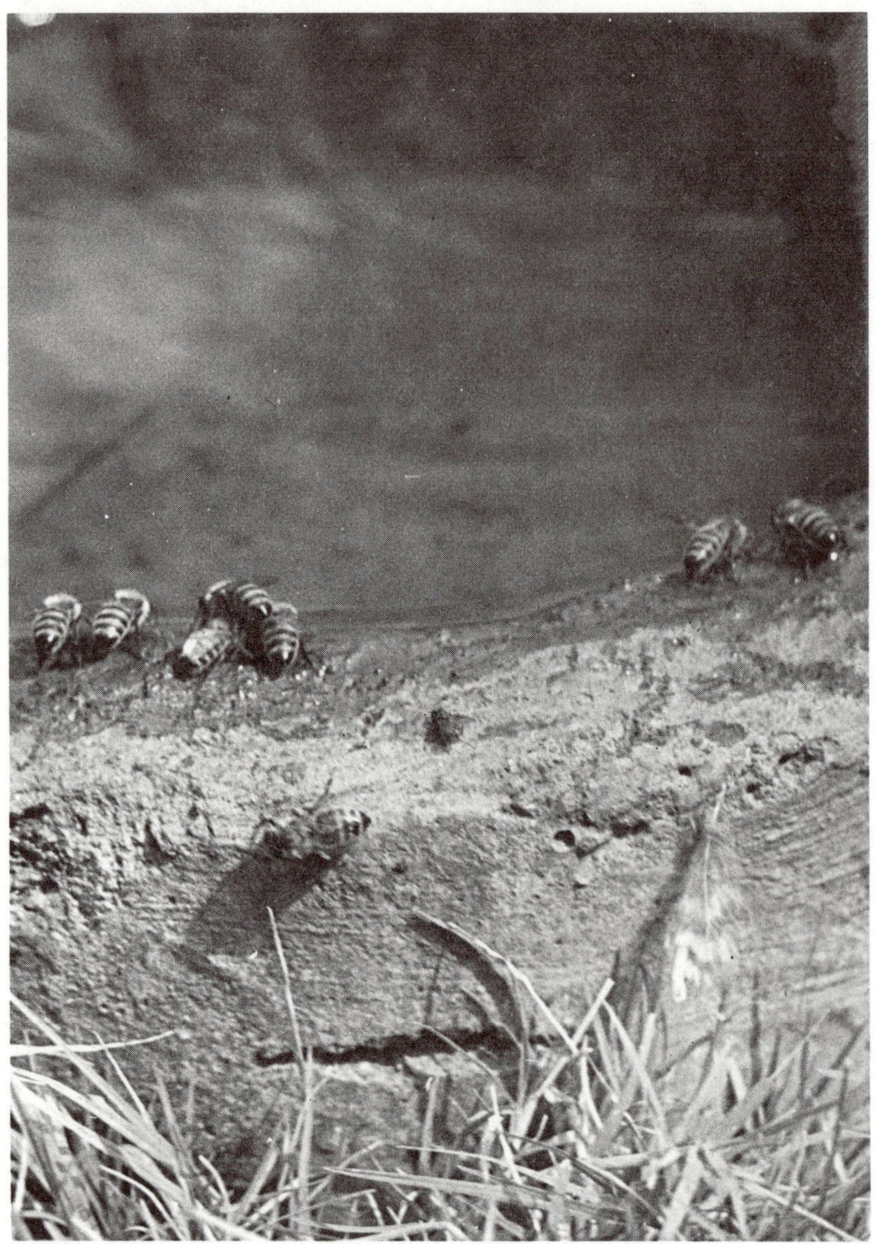

Fig. 5. Bees need a constant source of water to aid in cooling the hive and evaporating the excess liquids from fresh nectar. Always provide a continuous source of water such as a hydrant slowly dripping onto a board. Don't let bees get started using the neighboring dog's water dish or a swimming pool as their source of water.

CHAPTER VII
DAMAGES

A piece of broiled fish, and of an honeycomb
. . . . . . Bible

For every legal wrong there is a legal remedy. And it is fundamental in the law of damages that the injured party is entitled to compensation for the loss sustained. Where property is destroyed by a wrongful act, the owner is entitled to its money equivalent, and thereby to be put in as good position pecuniarily as if his property had not been destroyed.

The word "damages" has been defined throughout existence and from early times. The eminent legal scholar, Blackstone, defined it as "The money given a man by a jury as a compensation for some injury sustained." Webster, in his dictionary, claimed that damages were "The estimated reparation for detriment or injury sustained." And finally, the legal scholar Rutherford, said it is "every loss or diminution of what is a man's own, occasioned by the default of another."

At the outset it is well to point out that damages are not recoverable for every injury. This was observed in a Pennsylvania case:[1]

> "A farmer's wife could not recover damages for heart disability resulting from her fright and shock upon being chased by trespassing animal which did not touch her."

Unless an act is wrongful in the sense of being unlawful, it will not sustain a suit for damages.[2] On the other hand, where there is an invasion of a legal right without actual damages, "nominal" damages are recoverable, usually 6¢ to $1.00. The law infers some damage from the invasion of a right; and, if no evidence is given of any particular amount of loss, it declares the right by awarding nominal damages. This is illustrated in a Virginia proceeding:[3]

> "Whenever a party has violated the rights of another, the party whose rights have been so violated has a cause of action against the other for damages; but in the absence of proof of some actual damage resulting from the wrongdoer's act, nominal damages only can be recovered."

The same result follows where actual loss is sustained but the amount of the loss is too speculative to be proved with reasonable certainty.[4]

Contrasted to nominal damages, actual or compensatory, the terms being synonymous, are damages in satisfaction of, or in recompense for, loss or injury sustained. The injured party must show a legal injury and a perceptible resultant damage. "The wrong done and the injury sustained must bear to each other the relation of cause and effect."[5]

In other words, the injured party must not only show that he has sustained damages, but must show with reasonable certainty the extent of such damages. And he must also show that such damages are the natural and proximate result of the injury complained of.[6] A New York case was on point:[7]

> "Honey production depends on many things, including weather, crops, care and length of the hazardous life of the bee, which is limited in any event. This is not the proper measure of damage. The measure of damage was the value of the damage to the bee colonies, plus damage to honey, wax or other food in existence at the time, and used or destroyed as a result of negligence found to be a proximate cause."

The law awards actual or compensatory damages only where in the unlawful act there is an absence of intentional wrong or malice or the act is not oppressively or recklessly committed. If, however, the unlawful act was done in a malicious manner, and proved so, the courts may award additional damages.

These "add-on" damages are spoken of as exemplary, punitive or vindictive and, in reality, the terms are synonymous. Where the destruction of animals or an injurious act to them is committed in a willful manner and under circumstances of aggrevation showing a violent, reckless and lawless spirit, exemplary or punitive damages may be recovered. As where the wrongful act "is done with a bad motive, or with such gross negligence as to amount to positive misconduct, or in a manner so wanton or reckless as to manifest a willful disregard of the rights of others."[8] The Commonwealth of Virginia says that: [9]

> "Punitive or exemplary damages are allowable only where there is misconduct or actual malice, or such reckless or negligence as to evince a conscious disregard of the rights of others. They are allowed so much as compensation for injured party's loss, as to warn others and to punish the wrongdoer, or with such malice as to evince a spirit of malice or criminal indifference to civil obligations."

And in Florida, punitive damages may be recovered upon a showing to the jury of malice, wantonness, oppression or outrage.[10]

It has been stated that the purpose of punitive damages is to deter the wrongdoer from similar conduct in the future as well as to deter others from engaging in such conduct. It is a protection for society as much as a punishment for the wrongdoer.

Authorities have recognized that it is competent for parties entering into an agreement to avoid all future questions of damages which may result from a violation thereof, and to agree upon a definite sum as that which shall be paid to the party who alleges and establishes the violation. In such a case, the damages so filed are termed liquidated, stipulated or stated damages. A Federal ruling stated:[11]

> "Though liquidated damages clauses are recognized and enforced by the courts, when freely entered into by the parties to a contract, such damages clauses are to be narrowly construed."

It is important to distinguish between liquidated damages and a penalty — the latter may be unenforceable. Whether a given sum, agreed to be paid in case of the breach of a contract is to be regarded as liquidated damages or as a penalty, must depend on the particular case, regardless of the name by which the parties have called it. If the contract is for the doing of a single specific act, and there is no adequate means of determining from the contract or otherwise the precise damages which may result from its breach, the sum agreed on will generally be regarded as liquidated damages and not a penalty. However, when from the nature of the contract or the work to be performed it is difficult or impossible to ascertain the damages resulting from a breach, the sum stipulated will generally be regarded as a penalty.[12] The Supreme Court of Arkansas ruled on the liquidated damages-penalty issue, stating:[13]

> "this court has consistently maintained the principle that the intention of the parties is to be arrived at by a proper construction of the agreement made between them and that whether a perticular stipulation to pay a sum of money is to be treated as a penalty, or as an agreed ascertainment of damages, is to be determined by the contract, fairly construed, it being the duty of the court always, where the damages are uncertain and have been liquidated by an agreement, to enforce the contract."

In determining whether the sum named in an agreement is a penalty or liquidated damages, the following rules may be applied:—[14]

(a) The courts will not be guided by the name given to it by the parties.
(b) If the matter of the contract is of certain value, a sum in excess of that value is a penalty.
(c) If the matter is of uncertain value, the sum fixed is liquidated damages.
(d) If a debt is to be paid by installments, it is no penalty to make the whole debt due on nonpayment of an installment.
(e) If some terms of the contract are of certain value, and some are not, and the penalty is applied to a breach on any one of them, it is not recoverable as liquidated damages.

There has not yet been discovered any standard by which to measure in dollars and cents the value of physical pain and suffering. It is a matter which must be left to the judgment and discretion of an impartial jury or judge. A Virginia case averred that:[15]

> "a jury verdict will not be disturbed in a case of a physical injury because the verdict is too large or too small, unless in the light of all the evidence, it is manifestly so inadequate or so excessive as to show plainly that the verdict has resulted from one or both of the two courses; (1) the misconduct of the jury, or (2) the jury's misconception of the merits of the case insofar as they relate to the amount of damages."

In injuries to personal property, the rule for determining the amount of damages is to subtract the fair market value of the property immediately after the injury from the fair market value thereof immediately before the injury; the remainder, plus necessary reasonable expenses incurred, being the damages.[16] This viewpoint was exemplified in the State of Arizona:[17]

> "In action for damages to apiary when bees were poisoned, measure of damages was difference between market value of colonies of bees at time they were damaged and value after bees were killed, with reasonable expenses incurred in effort to mitigate loss."

And as early as 1904, in Texas, damages were allowed upon the fair market consideration calculated before and after injury. The court ruled:[18]

> "Where, prior to the shipment of bees, owner informed the carrier's agent that he had sold the bees at $3.65 per stand, delivered, and after the bees had been injured in transportation owner examined them and was familiar with their market value and the extent of their injury, he was entitled to testify as to the amount of damages sustained, less than the price for which the bees had been sold."

It is to be noted that the court looks to the market price, and nothing else, as the value of the injured property. An old axiom still holds good today: "The worth of a thing is the price it will bring." The law stresses that where there is no market value for certain kinds of property or where the market value does not represent the real value to the owner, testimony may be considered to show what such property is really worth in the business, and what a buyer would pay for it . . . . . but was not obligated to buy, and what one would take for such property in selling . . . . . who was not obligated to sell.[19]

A Minnesota court explained the meanings of direct and consequential damages, saying:[20]

> "The direct damage for fraud which includes a contract is the difference in value between what the party defrauded parted with and what he received. In addition to this the party defrauded may recover consequential damages flowing naturally and proximately from the breach. If one, through fraud, procures a sale of bees afflicted with disease, the purchaser may recover for the loss of other bees of his own to which the disease is communicated."

As a general rule damages, which are in their nature uncertain, speculative or contingent, cannot be recovered. As between possible methods by which a loss may be computed, the law prefers that which leads to certain, and not speculative results. A reason given for this rule is that uncertain or speculative damages are not susceptible of the exactness of proof that is required to fix a liability.[21] In New York State the rule states:[22]

> "Evidence as to the amount of honey which the bees would produce and the value thereof, was improperly admitted because such evidence was entirely too speculative."

Arizona had a similar ruling:[23]

> "In action for damages to apiary from poison, bee owner held not entitled to damages by reason of loss increase; such claim being too speculative."

And in Minnesota:[24]

> "But damages for loss of the subsequent increase of the diseased bees seems to us quite too remote to be recovered."

A court case in the State of Arizona included many of the points mentioned heretofore — these points are summarized from the decision of the judges:[25]

> "Stated in broad terms, however, the measure of damages is such sum as will compensate the person injured for the loss sustained, with the least burden to the wrongdoer consistent with the idea of fair compensation, and with the duty upon the person injured to exercise reasonable care to mitigate the injury, according to the opportunities that may fairly be or appear to be within his reach . . . . . The colony is a unit. The worker bees, the drones, and a queen together are valuable, but separate them and they are valueless.

So a colony or hive must be thought of and treated as a unit, just as a cow or a sheep is thought of and treated as a unit. As the owner may recover damages for the breaking of his cow's or sheep's leg, so may the beekeeper recover damages to his colony or hive caused by the killing or weakening of his bees. The killing or destroying of a few bees of a hive or the major portion thereof is an injury to the colony. Damages would be to the colony and not for the value of the bees as separate entities. We think the true measure of damages . . . . is the difference between the market value of the colonies at the time they were damaged and their value after they were rebuilt, together with the reasonable expenses incurred by plaintiff in an effort to mitigate or keep down the loss as much as possible. Of course, where the colonies were entirely lost or destroyed, plaintiff would be entitled as damages to the market value of them at the time of their loss or destruction. According to the rule stated . . . . there being some evidence to support such damages (please consider the case was decided in 1933):

The market value of 75 colonies destroyed or absorbed, $7.50 each or	$ 562.50
Damages to 308 rebuilt colonies at $4.50 each	1386.00
Wages of 2 apiculturists	100.00
Honey fed to bees	80.00
Wages of plaintiff for extra work caused by the poisoning of the bees	300.00
A total of	$2428.50

In addition to the above items, plaintiff claimed damages by reason of the loss of increase, fixed at 250 hives. This claim, it seems to us, is entirely too speculative and uncertain. While the bee is industrious, dependable, and intelligent, it is shortlived. Sixty to ninety days is his allotted time . . . . What colony or how many would have doubled and swarmed in the following spring is too much of a guess to be the basis of a claim for damages.

There was also a claim for loss of honey, but it was entirely without support in the evidence."

CITATIONS

1) — *Bosley v. Andrews,* 142 A.2d.263, 398 Pa.161
2) — *Pickens v. Coal River Boom, etc. Co.,* 41 S.E.400, 31 W.Va.445
3) — *Building Light, etc. Co. v. Fray,* 32 S.E.58, 96 Va.559
4) — *Bowen v. Fidelity Bank,* 209 N.E.140
5) — *Hoge v. Prince William Co-op Exch.,* 126 S.E.687, 141 Va.676
6) — *Diggs v. Lail,* 114 S.E.2d.743, 20 Va.871
7) — *Newton v. State,* 222 N.Y.S.2d.959, 31 Misc.2d.48
8) — *Bolten v. Gates,* 100 P.2d.145, 105 Colo.571
9) — *Giant of Va., Inc. v. Pigg,* 152 S.E.2d.271, 207 Va.679
10) — *Florida East Coast Railway Co. v. Cain,* Fla.App., 210 So.2d.481
11) — *United States v. Marietta Mfg. Co.,* S.D.W.Va., 53 F.R.D.390
12) — *Stony Creek Lumber Co. v. Fields,* 45 S.E.797, 102 Va.1
13) — *Blackwood v. Liebke,* 113 S.W.210, 87 Ark,545
14) — *Law of Contracts,* 2nd. Edition — Laurence P. Simpson, West Publishing Co., St. Paul, Minn.
15) — *Dalton v. Johnson,* 129 S.E.2d.647, 204 Va.102
16) — *Biederman v. Henderson,* 176 S.E.433, 115 W. Va. 374
17) — *S. A. Gerrard Co., Inc. v. Fricker,* Ariz.App., 27 P.2d.678

18) — *International & G.N.R. Co. v. Aten,* Tex.App., 81 S.W.346
19) — *Weber v. Wisconsin Power and Light Co.,* 255 N.W.261, 215 Wis.480
20) — *Sampson v. Penny,* 187 N.W.135, 151 Minn.411
21) — *Barnes v. Graham Virginia Quarries, Inc.,* 132 S.E.2d.395, 204 Va.414
22) — *Id,* 222 N.Y.S.2d.959, 31 Misc.2d.48
23) — *Id,* Ariz.App., 27 P.2d.678
24) — *Id,* 187 N.W.135, 151 Minn.411
25) — *Id,* Ariz.App., 27 P.2d.678

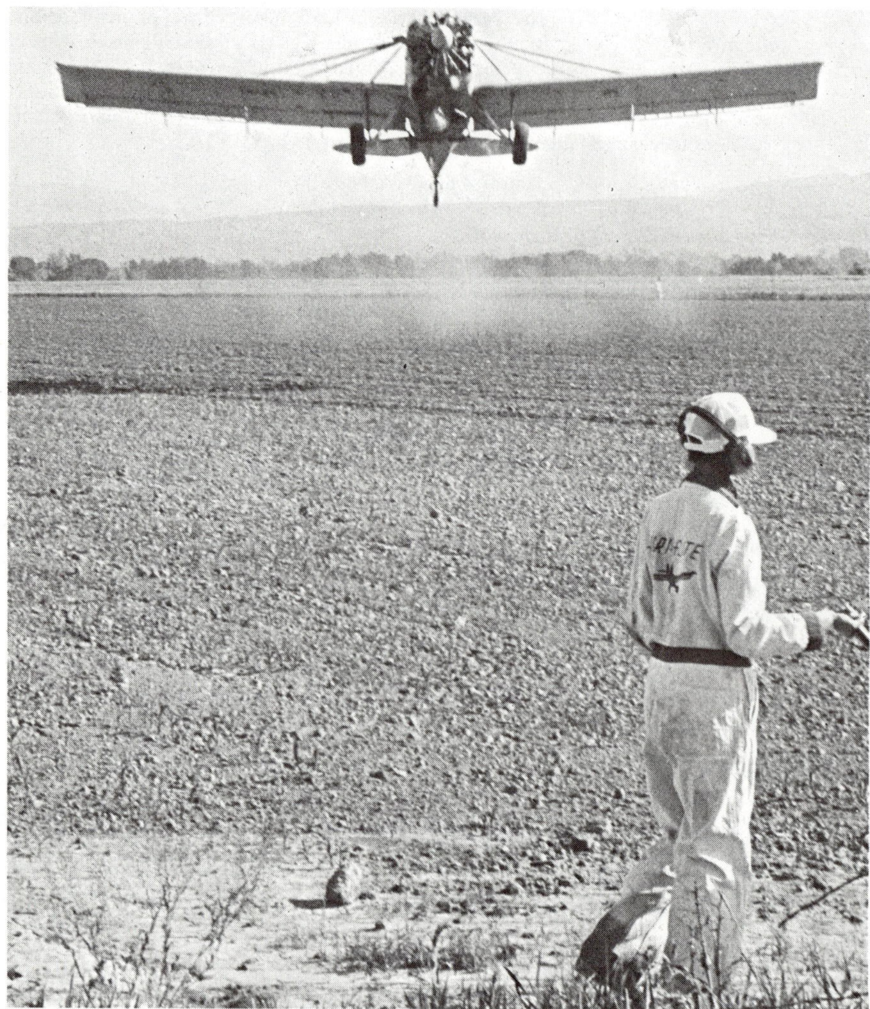

Fig. 6. A current record-keeping system will enable the beekeeper to quickly establish market value of his hives in case of pesticide damage. Keep up on the neighborhood crops. Make it a point to be aware of spraying and dusting plans so measures can be taken to confine the bees during the most dangerous period.

CHAPTER VIII

PESTICIDES

A bee-hive's hum shall soothe my ears
. . . . Samuel Rogers

Practically every species of plant has disease, insect, or weed pests which significantly influence its growth. The increased acreage in crops favors the development of insects, mite, and disease problems. This in turn usually requires that more pesticide treatment be made. Agricultural chemicals or pesticides for specific purposes are utilized to alleviate the countless pests of plants. Therefore, individual problems are controlled with chemicals for specific uses such as ascaricides, antibiotics, chemosterilants, defoliants, desiccants, fungicides, herbicides, insect growth regulators, insecticides, nematocides, and plant growth regulators.

Unfortunately, the honey bee is susceptible to many of the pesticides used in such an intensive plant control program, As a result, the honey bee is subjected to a continuous hazard of chemical poisoning that overshadows all other problems including bee diseases.

In collecting nectar and pollen, the honey bees contact many pesticides deposited upon the plants. The injurious chemicals settle on plants by the:[1]

a) drifting of toxic sprays
b) contamination of flowering cover crops when orchards are sprayed
c) dusting of crops

With few exceptions, pesticides applied as dusts are more hazardous to honey bees than those applied as sprays. And pesticide applications by aircraft over bees in flight are more hazardous than applications by ground equipment. The safest method of pesticide treatment is by granular application.

Symptoms of pesticide poisoning are somewhat similar to symptoms caused by plant poisioning or by adult bee diseases. The most positive indication of serious pesticide poisoning is the appearance of massive numbers of dying and dead bees in the colony entrances throughout the apiary. Pesticide kills may also be suspected when the weather has been warm, dry and pleasant for any length of time and the colonies are located in blooming crop and forage areas. A knowledge of local pesticide programs is important to avoid poisoning.

Affected bees usually are restless and are found crawling out of the hive, often unable to fly except for short distances. Many are trembling or crawling and tumbling about aimlessly. Many others remain in the hive by other hive workers. The legs of the affected workers may be dragged along as if paralyzed and the rear wings may be unhooked from the front wings and held at abnormal angles.

In addition to the 15% of colonies killed outright, the remaining colonies are weakened to such extent that they are no longer productive or effective pollinators, nor can they be used for divides to increase colony numbers. The monetary loss to beekeepers is large but the value of seed and fruit loss through lack of pollinators is 50 to 100 times greater.

More recently, there have been numerous reports of beekeepers using aerosol space spraying devices in their bee houses. These devices dispense pesticides such as lindane or DDVP. Also NO PEST STRIPS containing Vapona have become popular for controlling insects in buildings. They are highly effective in controlling the insects and are safe to people; however, they give off fumes which are readily absorbed into beeswax and will kill all bees placed in the contaminated equipment for several months.

The federal and state regulations governing the sale and use of pesticides are designed to protect the producer as well as the consumer from the misuse of poisonous materials. Generally, these are governed by health considerations for man, domestic animals, and wild life. Beekeeping is given a considerable measure of protection by these regulations, but the use of pesticides still remains an ever-present hazard to the beekeeping industry.

Beekeepers can reduce their losses by becoming familiar with the chemical control programs of the regions in which they operate their colonies. It is also helpful when beekeepers register the number of hives and their locations with the county regulatory officials who have authority over the use of pesticides, and request notification in advance of the use of chemical compounds that are likely to be highly injurious to their bees. With notification, the beekeeper is aware that his bees may be poisoned and he can help them recover.

Some states have laws or regulations which require that before a pesticide application toxic to bees can be made the owner of the bees within one (1) mile of the treated area must be notified and the beekeeper is given 48 hours to move or protect his bees.

It has been shown that spraying or dusting made at night, early morning, or in late afternoon, when the pollinators are not active, results in less damage than if made during the hour of major flight of bees. Above all, certain highly toxic pesticides should not be applied to fruit trees, legumes, or other crops while they are in bloom.[2]

In California, as in other states, bee laws provide the basis for an effective apiary inspection program to help beekeepers protect honey bees from disease, pesticides and theft. In addition, beekeeping in some localities is governed by city or county ordinances so beekeepers should consult local authorities.

All apiaries, in California, must be registered each November 1 with the Agricultural Commissioner of the county in which bee colonies are located. Registration is free and consists of listing the location of each apiary and the number of colonies at each location. Newly acquired colonies and colonies brought into the state must be registered within thirty (30) days of establishment. If the beekeeper changes the location of an apiary, he must inform the County Agricultural Commissioner within five (5) days of the change of location. He must identify each apiary, in a conspicuous place near the entrance side of the apiary, using a sign stating in black letters . . . . . the owner's name, address and phone number. If the beekeeper wishes to be notified of pesticide applications within one (1) mile radius of his apiary, he must submit a written report to the County Agricultural Commissioner requesting notifications. In some counties, the beekeeper may make a choice of the pesticides or groups of pesticides for which he wishes to be notified. If the beekeeper has not complied with these requirements, he may not be able to recover losses due to pesticide applications.[3]

The California regulations, mentioned heretobefore, were put to a rigid test; a beekeeper sued a landowner who had applied a pesticide, parathion, to his almond orchard. The spray destroyed many bee colonies which were pollinating the orchard. The beekeeper failed to comply with the registration requirements of the California Code. However, the orchard owner knew of the presence of beekeeper's apiaries and knew how to contact him. The judges of the California Court of Appeal ruled that:[4]

> "The regulations and the statutes relied upon by the orchard owners are fair and strike a reasonable balance between the respective agricultural interests involved. (Notice by the bee owner to the county commissioner of the location or relocation of his bee colonies.) However, we believe that beekeeper's failure to give notice does not bar his action against almond growers. Here almond growers, engaged in pest control, knew of the presence of the apiaries. The primary function of the statutes and regulations cited, to furnish *notice* to apiarists and pest control operators, had been accomplished. The orchard owners had notice, admitted they were aware of the presence of the apiaries, yet proceeded to apply parathion without notifying the beekeeper and without regard for the welfare of the apiaries. We believe that section 29241 was not intended as an unwarranted shield for the avoidance of a just duty of care, which is the purpose for which orchard owners seek to employ it."

CITATIONS

1) — *Division of Agricultural Sciences,* U. of Calif, Leaflet 2286

2) — *E. Laurence Atkins,* Injury to Honey Bees by Poisoning, The Hive and the Honey Bee, A Dadant Publication.

3) — *Id,* Injury to Honey Bees by Poisoning, p.686

4) — *Hall v. C & A Navarra Ranch, Inc.,* 101 Cal.Reptr.249, 24 Cal.App.2d.774

Fig. 7. Honey bees are one of Nature's most valuable insects because of their pollination services. Fruit growers need bees to pollinate the blossoms and beekeepers benefit from the abundance of nectar and pollen.

CHAPTER IX
TRESPASS

All Nature seems at work, Slugs leave their lair -
The bees are stirring - birds are on the wing -
. . . Samuel Taylor Coleridge

Simply stated, trespass is an entry onto somebody else's property without legal authority or special justification. Bees, by trespassing, may cause annoyance, interfere with the use or enjoyment of property, or lower the value of property.

On the other hand a comment from the Restatement has a different view, stating:[1]

> "It has also been stated that bees are among those usually harmless domestic animals which cannot be constantly confined without interfering with their economic usefulness, and that, accordingly, in certain locations they have been allowed as a matter of tradition to run at large without being viewed in the same light as livestock, for whose entry upon another's property, their keeper would be strictly liable in trespass."

The law has undergone considerable change on the question of trespass. Formerly, the owner of property — a slave, an animal, even an inanimate thing — was generally held strictly liable for the harm it did. He was so totally identified with his property that the law held him liable, without being at fault himself, for the damage his property inflicted on his neighbors. It was once said: "Where my animals of their own wrong, without my will and knowledge, break another's close (enter onto another's private property), I shall be punished, for I am the trespasser with my animals — for I am held by the law to keep my animals without their doing wrong to anyone."

This primitive notion of absolute liability is no longer with us. The law now recognizes that the degree of liability can differ according to the difference of animals, the circumstances of the case.

Our courts, however, appear to be at a variance as to the liability of the bee as a trespasser. In Delaware:[2]

> "Trespass quare clausum fregit (common-law remedy for the recovery of damages for the wrong of intruding upon the real propery of another) is an inappropriate remedy to recover for injuries done by bees to the person or property of another."

Conversely in Vermont:[3]

> "Under the regulation of property rights since the institution of civil society, the forest, as well as the cultivated field, belongs to the owner thereof, and he who invades is a trespasser."

A recent California proceeding, relative to pesticides, ruled on the consequences of trespassing bees:[4]

> "If bees procured a poisonous compound from which they died while they were trespassing on the fields of other owners land, it appears that the

> apiarist could not recover damages unless the poison was distributed wantonly, maliciously, or with the deliberate intent to injure or destroy the bees. . . . there is no duty on the owner of the land to protect trespassing bees from the danger of a poisonous compound sprayed or dusted on his land."

In many foreign countries bees are not considered trespassers in their pollinating flights. However, the apiarist must prove that the trespassing bees are his property. In England, the bee owner can only recover them at the risk of committing trespass, if permission is refused to enter the land.

Our courts, ruling on trespassing bees, would be wise to take notice of the economic value of the honey bee. A pamphlet, distributed by the New York State College of Agriculture and Life Sciences, would aid the eminent judges in their deliberations. It reads:[5]

> "The pollination of agriculture crops is the most important contribution of honey bees to our national economy. Although the value of honey bees for pollination cannot be estimated, it is many times the total value of both honey and beeswax which they produce. Without cross-pollination many crops would not set seed or produce fruit. Many insects other than the honey bee can carry pollen from one plant to another; but in areas where agriculture has been intensified the number of these other insects is inadequate for commercial pollination."

In this chapter the issue of trespass confines itself with the bee as trespasser. Additional material, in a like vein, can be found in the chapter on Nuisance. The chapters Negligence and Contributory Negligence have cases of trespass where the person is the trespasser.

CITATIONS

1) — *Restatement, Second Torts,* comment
2) — *Petey Mfg. Co. v. Dryden,* Del. App., 62 A.1056
3) — *Adams v. Burton,* 43 Vt.36
4) — *Lenk V. Spezia,* 213 P.2d.47, 95 Cal.2d.296
5) — *Morse, R.A. and Dyce, E.J.,* Beekeeping; General Information

CHAPTER X
LARCENY

But for your words, they rob the Hybla bees,
And leave them honeyless . . . .
Shakespeare

In legal terminology, the taking of personal property accomplished by fraud or stealth, with intent to deprive another thereof, is larceny.[1] It is the unlawful taking of one's lawful property. As stated in an early Virginia case :[2]

> "Although ferae naturae, bees in the possession of any person are the subject of larceny. For although not fit for food themselves, their honey is, and for that reason they are deemed to be within the common-law rule that animals ferae naturae, when reclaimed or confined, are the subject of larceny if fit for food."

Similarly, the State of Indiana ruled:[3]

> "The indictment is not objectionable on account of the charge in it respecting the bees. They are not only alleged to be the goods of the Chapmans, but to have been in their possession when they were stolen. Bees are, no doubt, ferăe naturae, but when they are in the possession of any person they are the subject of larceny."

These century old cases hold true today as in the past years. Presently, larceny is more sophisticated. Not only are bees and honey pilfered but wax, frames, hive bodies and any other apiary item of value. Price increases of beekeeping equipment seems to lure some to this unsavory practice.

Larceny, for one reason or another, is often confused with robbery. But there is a distinction. Robbery is the unlawful taking of personal property in the possession of another, from his person or immediate presence, and against his will, accomplished by means of force or fear.[4] Larceny has been defined heretobefore.

Both offenses are statutory and the punishment is invariably set by law. Only of late have those committing larceny of bees, honey, and equipment been receiving the punishment they deserve.

In past years the bee felons were often set free or received little punishment. The reasons — many and varied: Prosecutors failed to appreciate the value of hives and bees or the seriousness of the crime; they steered their time to more serious crimes, in their opinion, and devoted little and unprepared time to the bee thief, after all, stealing bees was a crime of non-violence; they had no idea how to prove real ownership of bees, hives and equipment; and they, in some instances, did not choose to work extra hard on criminal cases that were not receiving considerable public attention.[5]

But times and conditions have changed.

Prosecutors now appreciate the value of hives and bees and the loss of income to the apiarist; they consider the larceny of bees, honey and equipment as a serious crime; they enlist the aid of state bee inspectors and bee association members, as well as check brand marks, for identification and ownership; and they receive their share of publicity as bees are newsworthy.

The vigorous prosecution of those charged with larceny of bees and apiary equipment has resulted in heavy fines and imprisonment. In California, a Sutter County court imposed a fine of $1500 and a 60 day jail term for one receiving stolen bee equipment. And in Bakersfield, California, a convicted felon was sentenced to six months in jail and placed on probation for three years.

Some states, among them Connecticut, Kentucky, and Nebraska, have statutes especially applicable to honey bees and the crime of larceny.

The taking of honey, produced by wild bees which have never been reduced to possession, does not constitute larceny. This was confirmed in New York where:[6]

> "The taking of honey produced in a tree on another's property by wild bees which had never been reduced to possession did not constitute 'larceny'."

However, if the tree on which these bees produce honey is on another's property, the taker may be charged with a crime — the crime of trespass, not larceny (if permission was not granted by landowner). This is stated in Vermont:[7]

> "The forest, as well as the cultivated field, belongs to the owner thereof, and he who invades it is a trespasser."

CITATIONS

1) — *State v. Ugland,* 187 N.W.237, 48 N.D.841
2) — *Harvey v. Commonwealth,* 23 Gratt. (Va.) 941
3) — *State v. Murphy,* 8 Blackf. (Ind.) 498
4) — *McDaniel v. State,* 16 Miss.401
5) — *Nelson E. Bailey,* American Bee Journal, Nov. '75
6) — *People v. Hutchinson et als.,* 9 N.Y.S.2d.656, 169 Misc.724
7) — *Adams v. Burton,* 43 Vt.36

CHAPTER XI
ZONING

No shade, no shine, no butterflies, no bees,
No fruits, no flowers, no leaves, no birds, —
November!

. . . . . . *Thomas Hood*

An energetic and spry senior lady of 92 years, a beekeeper for 58 of those years, attended the council meeting in her city. She went to city hall expecting to lose the right to operate her bee and honey business — revoke her conditional use permit. Fortunately, the council members agreed to preserve her conditional use permit to keep nine hives on her one acre fruit and honey orchard.[1]

Many beekeepers, however, are not as fortunate as the 92 year-old lady. Some are required to relocate their hives, some have to reduce the number of bees, and some just "close shop." The reason — zoning.

Legally, zoning is the "division of a municipality or other local community into districts, and the regulation of buildings and structures according to their construction and the nature and extent of their use, or the regulation of land according to its nature and uses."[2]

In a sense, zoning is separating commercial or industrial districts of a city or town from the residential districts, and prohibiting the establishment of places of business or industry in any district designated as residential.

It is to be understood that zoning ordinances which prohibit or restrict beekeeping in a city or town are legal. Just as it is legal for a city to prohibit gasoline stations in residential areas, it is also lawful for a city to prohibit beehive locations in specific areas of the city.

Zoning regulations of land use can only be imposed where they bear some reasonable relation to the public's health, general welfare, morals or safety. These regulations or ordinances, prohibiting placement of beehives in certain areas of a city or town, have been held by the courts to be reasonably related to the public's health and safety, and therefore, the courts have held such ordinances to be lawful. Also, ordinances requiring beehives to be located a certain number of feet away from property lines, or a certain number of feet from houses and buildings, have been held by the courts to be legal.[3] A California court followed this train of reasoning, saying:[4]

> "The beekeeper questions the reasonableness of the regulation . . . . . . and she contends that to permit beekeeping in such sections and to prohibit it in the area where the beekeeper resides is an unreasonable, arbitrary and unlawful discrimination in the exercise of the police power.
>
> The beekeeper lays great stress on those advantages generally and stresses particularly the benefits to the residents in her community resulting from the cross-pollination of the fruit blossoms and flowers in addition to the commercial value of the bees. These advantages may be conceded.

> Nevertheless, municipalities may regulate the use to be made of property in the interest of the public health and welfare, so long as the attempted regulation is not unreasonable or arbitrary."

In 1957, the R-1 zone of the Los Angeles Municipal Code stated: The R-1 zone may be used in . . . "the raising of poultry, rabbits and chinchillas and the keeping of domestic animals in conjunction with the residential use of a lot . . . ." The beekeeper claims that his bees are domestic animals within the meaning of the section in question. In its ruling the Superior Court disagreed with the beekeeper:[5]

> "Bees were not domestic animals within meaning of municipal code section authorizing the keeping of domestic animals in conjunction with residential use of lots in a certain zone."

As a general rule, zoning regulations do not prohibit the right to continue such uses of land as were in existence at the time of adoption of the new regulation. Sometimes zoning regulations expressly provide that nonconforming uses may be continued, at least for a specified period of time.[6]

It must be noted that to continue a nonconforming use of property applies only to a lawful nonconforming use which existed at the time of promulgation of the zoning regulation. If the apiary is considered a nuisance, labeling it as nonconforming will be a mere redundancy.

The apiary, to preserve its nonconforming use, is to continue only the same use of the property as existed prior to the date of the zoning regulation. It may not change to a different kind of nonconforming use except when such change is permitted by the terms of the zoning regulation.

In most jurisdictions a substantial enlargement or extension of a nonconforming use ordinarily is not permitted. But this enlargement or extension does not prevent an increase in the amount of use within the same area. It was held in a New York Court that:[7]

> "a landowner has a right to accommodate an increase in business and to erect buildings necessary thereto. If the business in essence is the same as the nonconforming use which existed prior to adoption of a zoning regulation."

Most courts impose the requirement of actual use, as distinguished from mere planned or intended use. A mere contemplated use is insufficient to establish an existing nonconforming use.

Some courts hold that a use of land may be maintained and continued as a nonconforming use only if it is a "substantial one," the destruction or elimination of which would cause "serious financial harm" or "peculiar hardship" to the property owner. It has accordingly been held that a property owner's pre-existing use of premises for harboring pigeons as a hobby did not amount to such a vested property right that it would be exempt from the prohibition of zoning ordinance.[8]

Ordinances commonly restrict the repair or alteration of nonconforming buildings and hives destroyed or partially destroyed by fire, hurricane, or other calamity. However, the provisions of these restrictions vary from place to place. The percentage of destruction beyond which rebuilding is prohibited averages sixty percent, but may be as high as seventy-five percent or as low as twenty percent. A New Jersey court ruled:[9]

> "a structure devoted to a nonconforming use may be restored or repaired in the event of partial destruction. What constitutes partial depends on the facts of each case, and the town zoning ordinance limiting the right to restore to cases of destruction of less than 75 percent of replacement values."

It is held that if a nonconforming use existing at the date of promulgation of a zoning ordinance has been permanently discontinued or abandoned, the right to such use is lost and compliance must thereafter be had with the zoning regulations. The State of Connecticut offers an explanation of "abandonment":[10]

> "The landowner, having for a substantial period voluntarily ceased to use his property for nonconforming purposes and having devoted the use to residential purposes, cannot now avail himself of the exemption accorded him and may not use the premises for or revert to the nonconforming use which was existing at the time when the zoning regulations were adopted."

It was generally held that the right to continue a nonconforming use is not limited to the owner of the property at the time the ordinance was enacted but extends to subsequent purchasers.

This generality, however, is slowly but surely falling by the wayside. More and more communities are following and enacting ordinances as that of Chicago:[11]

> "if the structure is not designed for the nonconforming use, the use must be discontinued upon transfer of ownership or termination of the existing lease."

In previous years, a lawful bee operation, existing prior to enactment of a zoning ordinance or regulation, had a vested right to continue its operation. This right is slowly being eroded. The theory of "amortization", relating to nonconforming uses, is creeping into ordinances and court decisions, offsetting previous thinking. A California court aptly describes amortization:[12]

> "The elimination of existing uses within a reasonable time does not amount to a taking of property, nor does it necessarily restrict the use of property so that it cannot be used for any reasonable purpose. Use of a reasonable amortization scheme provides an equitable means of reconciliation of the conflicting interests in satisfaction of due process requirements."

In Wichita, nonconforming commercial and industrial buildings in residence areas are given sixty years to move out. In Dallas, amortization was worked out by requiring the use be discontinued at the expiration of five years.[13]

City planners and beautification experts deem it necessary to gradually eliminate nonconforming uses as areas become populated and urbanized. A 1964 law review article pointedly sums up amortization:[14]

> "nonconforming uses can and must be eliminated. It is suggested that the amortization scheme, by balancing the benefit gained by the public against the loss sustained by the individual loser, is the fairest method of achieving this goal."

The dictionary defines variance as something different, a deviation. In zoning, variance is "an exception from the application of a zoning regulation granted by proper authority to relieve against practical difficulties and unnecessary hardship."[15] It should be noted that financial hardship, by itself, may not be sufficient to acquire a variance.

Another zoning issue, that may be of interest to the apiarist, is "spot zoning". This is "a carving out of one or more properties located in a given use district and reclassifying them in a different use district."[16] A Connecticut court elaborated on spot zoning, ruling:[17]

> "it is a provision in a zoning plan or a modificatin in such plan, which affects only the use of a particular piece of property or a small group of adjoining properties and is not related to the general plan for the community as a whole."

CITATIONS

1) — *The Register,* Fullerton, Orange County, Calif.,11/29/79
2) — *Angermeier v. Sea Girt,* 142 A.2d.624, 27 N.J.298
3) — *Nelson E. Bailey,* American Bee Journal, January 1976
4) — *Ex parte Ellis,* 81 P.2d.911, 11 Cal.2d.571
5) — *People v. Kasold,* Calif. App., 314 P.2d.241
6) — *Id,* American Bee Journal, January 1976
7) — *Gerling v. Bd. of Zoning,* 167 N.Y.S.2d.358
8) — *People v. Miller,* 106 N.E.2d.34, 304 N.Y.105
9) — *H.Behlen & Bros. v. Town of Kearny,* 105 A.2d.894
10) — *Town of Darien v. Webb,* 162 A. 690
11) — *City of Chicago, Illinois*
12) — *City of Los Angeles v. Gage,* 274 P.2d.34
13) — *35 Va.L.Rev.348*
14) — *Md.L.Rev.2d.323*
15) — *58 Am.J.1st.* Zon. Sect. 194 et seq.
16) — *Chayt v. Maryland Jockey Club,* 18 A.2d.856, 179 Md.390
17) — *Eden v. Town Plan and Zoning Comm. of Town of Bloomfield,* 80 A.2d.746, 139 Conn.59

CHAPTER XII

HONEY — RULES AND REGULATIONS

Isn't it funny
How a bear likes honey?
. . . *Alan Alexander Milne*

Many outstanding texts have excellent descriptions of the characteristics, kinds, physical and chemical properties, colors, flavor, aroma, and compositions of honey. For further information consult these books for a more detailed discussion of the particular subject.

According to the United States Food and Drug Administration (informal description), "Honey is the nectar and saccharine exudation of plants, gathered, modified and stored in the comb of honeybees (Apis mellifera and Apis Dorsata); is levorotatory, contains not more than twenty-five (25%) per centum of water, not more than twenty-five hundredths (0.25%) per centum ash, not more than eight (8%) per centum of sucrose. The various states have a like definition as that of the Food and Drug Administration though some vary to an extent, as to the per centums.

Honey rules and regulations are varied and many, however, for the discussion here, the classes, standards, label requirements, foreign importation, and loan and purchase program, will be outlined and discussed.

Plant source names are usually the means of identification of honey, such as alfalfa honey, sage honey, clover honey, etc. The Food and Drug Administration, by regulation, has stated that honey cannot be labeled from the plant or blossom source except where the plant or blossom is the chief floral source of the product.

*The Agricultural Marketing Service of the United States Department of Agriculture has issued standards for grades of extracted honey as well as comb honey. This system of classification is by methods of market production and preparation:

1. Extracted honey is honey that has been separated from the comb by centrifugal force, gravity, straining, or by other means. It may be processed and prepared for market as one of the following types:

(a) Liquid honey — free of visible crystals.

---

*United States Standards for grades of Extracted Honey, effective April 16, 1951, and United States Standards for grades of Comb Honey, second issue, as amended, effective May 23, 1967, may be obtained from:

Chief, Processed Products Standardization and Inspection Branch
Fruit and Vegetable Division, AMS
United States Department of Agriculture
Washington, D.C. 20250

(b) Crystallized honey — solidly granulated or crystallized, irrespective of whether "candied", "fondant", "creamed", or "spread" types.

(c) Partially crystallized honey — mixture of liquid honey and crystallized honey.

2. Comb honey — contained in cells of the comb and appears on the market as:

(a) Section comb honey — produced in squares of 4-1/4 x 4-1/4 x 1-7/8 inches or rectangles of 4 x 5 x 1-3/4 inches. May also appear in circular sections.

(b) Individual section comb honey — small sections, usually one fourth the size of ordinary sections.

(c) Bulk comb honey — produced in shallow extracting frames and usually sold when filled as complete units.

(d) Cut-Comb honey — bulk comb honey cut into pieces of different sizes, edges drained or extracted, and the individual pieces wrapped for market.

(e) Chunk honey — cut-comb honey packed in containers filled with liquid extracted honey.

The grades of extracted honey are:

(a) U.S. Grade A or U.S. Fancy — contains not less than 81.4% soluable solids, possess a good flavor for the predominant floral source or, when blended, a good flavor for the blend of floral sources, is free of defects, and is of such quality with respect to clarity as to score not less than 90 points.

(b) U.S. Grade B or U.S. Choice — contains not less than 81.4% soluable solids, possess a reasonably good flavor for the blend of floral sources, is reasonably free from defects, is reasonably clear, and scores not less than 80 points.

(c) U.S. Grade C or U.S. Standard — reprocessing honey that contains not less than 80% soluable solids, possess a fairly good flavor for the predominant floral sources, or, when blended, a fairly good flavor for the blend of floral sources, is fairly free of defects and is of such quality with respect to clarity as to score not less than 70 points.

(d) U.S. Grade D or Substandard — fails to meet requirements of U.S. Grade C or U.S. Standard.

The factors as to the calculation of the point score are evaluated as:

Flavor	50 points
Absence of defects	40 points
Clarity	10 points
Total score	100 points

The grades of the variations of comb honey are:

1. Comb-section honey
   a) U.S. Fancy
   b) U.S. No. 1
   c) U.S. No. 2
   d) Unclassified
2. Shallow-frame comb honey
   a) U.S. Fancy
   b) U.S. No. 1
   c) Unclassified
3. Cut-comb (wrapped) honey
   a) U.S. Fancy
   b) U.S. No. 1
   c) Unclassified
4. Chunk or bulk comb honey — packed in tin
   a) U.S. Fancy
   b) U.S. No. 1
   c) Unclassified
5. Chunk or bulk comb honey — packed in glass (no given volume is required)
   a) U.S. Fancy
   b) U.S. No 1
   c) Unclassified

Now that honey has been standardized and classified, label inscriptions for the honey container must conform to requirements of the appropriate agencies. A synopsis of label requirements for labeling honey follows:*

A. Bulk Honey in Drums, Gallons, Larger Container Sizes:

1. Brand Name, If Any. (see paragraph 7(b) below)
2. Common or Usual Name of Food: HONEY

   (a) In bold type on the principal display panel — directly on drum or other container or on label affixed to container.

   (b) Size of type shall be reasonably related to the most prominent printed matter on the label.

   (c) The name "Honey" shall be on a line generally parallel to the base on which the container rests.
3. Name and Address of Producer, Packer or Distributor.

   (a) If distributor, name must be qualified by phrase such as "Packed for", or "Distributed by", before the name, or "Distributor" following the name.

---

*As prepared by Robert M. Rubenstein, Counsel of the Honey Industry Council, New York, New York, in Gleanings of Bee Culture, June, 1977.

(b) *Street address* of producer, packer or distributor not necessary *if* place of business is shown in a current city or telephone directory.

(c) Zip code must appear in address.

(d) Where producer or packer packs honey in places other than its principal place of business, it may use the address of its principal place of business in lieu of the actual place of packing or distributing.

4. Declaration of Net Quantity of Contents.

(a) The statement of contents shall be in terms of weight (avoirdupois pound) unless there is a firmly established general trade custom of declaring the contents of honey by fluid measure in which case fluid measure may be used. The statement of contents shall appear on the principal display panel (which may be the label).

(b) The statement shall be in conspicuous and easily legible bold face print or type in distinct contrast (by typography, layout, color, embossing, or molding) to other matter on the label. The letters may be no more than 3 times as high as they are wide.

(c) The statement shall include the words "net weight" or "net wt.".

(d) If desired, (but not required), an *additional* weight statement in terms of the metric system may appear on the principal panel or elsewhere on the container.

5. Ingredient Statement — Not required if product is solely honey. If product is a mixture or blend of honey with any other substance or substances, the ingredients must be listed by common or usual names, in order of decreasing predominance. This ingredient statement may appear on any appropriate part of the label but the entire statement must appear on a single panel of the label.

6. Exemption from Labeling Requirements.

(a) Shipments of bulk honey in interstate commerce are exempt from above labeling requirements *if* (and only if) either of the following two conditions exist:

(1) The shipper is the operator of the establishment where the honey is to be processsed, labeled, or repacked; or

(2) In case the shipper is not such operator, such shipment or delivery is made to such establishment under a written agreement, signed by and containing the Post Office Address of the shipper and such operator and containing such specifications for the processing, labeling, or repacking, as the case may be, if such specifications are followed, that such food

will not be adulterated or misbranded upon completion of such processing, labeling, or repacking.

Both shipper and such operator must each keep a copy of such agreement 2 years after final shipment of the honey from such establishment. Also must allow FDA inspectors to examine these copies.

7. Illegal Representation on a Label.

(a) Labelling a *mixture* of honey and another food (e.g. corn syrup) and calling it "Honey" even though the other ingredient is listed in an ingredient statement.

(b) Any representation on the label that expresses or implies a geographical origin of the honey unless that representation is a truthful representation of geographical origin, is a trademark or a trade name which has so long and exclusively been used by a packer or distributor of that honey that it is generally understood by the consumer to mean the product of a particular packer or distributor or is so arbitrary or fanciful that it is not generally understood by the consumer to suggest geographic origin.

B. Consumer Size Packages (Under 4 Lbs. or Under 1 Gal.)

Label must contain:

1. Brand Name, If Any.(see paragraph 7 (b) below)

2. Common or Usual Name of Food: HONEY

(a) Name in bold type, on principal display panel — directly on jars or other container, or on label affixed to container.

(b) Size of type shall be reasonably related to the most prominent printed matter on the label.

(c) The name "Honey" shall be on a line generally parallel to the base on which the container rests.

3. Name and Address of Producer, Packer, or Distributor.

(a) If distributor, name must be qualified by a phrase such as "Packed for", or "Distributed by", *before* the name, or "Distributor" *following* the name.

(b) *Street Address* of producer, packer or distributor not necessary *if* the place of business is shown in a current city or telephone directory.

(c) Zip code must appear in the address.

(d) Where the producer or packer packs honey in places other than its principal place of business, it may use the address of its principal place of business in lieu of the actual place of packing or distributing.

4. Declaration of Net Quantity of Contents.

(a) The statement of contents shall be in terms of weight (avoirdupois pounds) unless there is a firmly established general trade custom of declaring the contents of honey by fluid measure in which case fluid measure may be used. The statement of contents shall appear on the principal display panel (which may be the label). The term "principal display panel" means the part of the label that is most likely to be displayed, or examined under customary conditions of display for retail sale.

(b) To determine the proper type size to be used in declaring the quantity of contents, the term "area of the principal display panel" must be taken into consideration. In the case of a cylindrical or nearly cylindrical jar, the area considered to be 40% of the height of the jar times its circumference; exclude shoulders and necks of bottles or jars.

(c) The declaration of net quantity of contents must be placed on the principal display panel *within the bottom 30% of the area of the label panel,* in lines generally parallel to the base of the package. The contents declaration shall be separated from other printed label information appearing above or below the declaration by a space equal to at least the height of the lettering used in the declaration and by a space equal to twice the width of the letter "N" of the style of type used in the quantity of contents statement, from other printed label information appearing to the left or right of the declaration.

On packages having a principal display panel of 5 square inches or less, the requirement for placement within the bottom 30% of the area of the label panel shall not apply, when the declaration of the net quantity of contents meets the other requirements outlined herein.

(d) The declaration of contents must be not less than 1/16th inch in height on jars or bottles where the principal display panel has an area of five square inches or less; not less than 1/8th inch in height on jars or bottles having a principal display panel area of more than 5 but not more than 25 square inches; not less than 3/16 inch in height on jars or bottles with a principal display panel of more than 25 but not more than 100 square inches only, and not less than 1/4 inch in height on jars or bottles having a principal display panel area of more than 100 square inches, and not less than 1/2 inch in height, if the area is more than 400 square inches. (Reminder: The size of the label is not synonymous with the size of the principal display panel. The label is usually smaller than the principal display panel).

If the declaration is blown, embossed or molded on a glass or plastic surface rather than by printing, typing or coloring, the lettering sizes above mentioned shall be increased by 1/16th inch.

(e) The declaration of contents must be expressed both in ounces, with identification by weight or by liquid measure, and, if one pound or one pint or more, must be followed in parenthesis by a declaration in pounds for weight units with any remainder in terms of ounces or common or decimal fractions of the pound, (e.g. 1-1/2 lb. weight shall be expressed as "Net wt. 24 oz. (1 lb. 8 oz.)", Net wt. 24 oz. (1-1/2 lb.)" or "Net wt. 24 oz. (1.5 lb.)". A declaration of less than one pound avoirdupois weight shall be expressed in ounces only. Abbreviations are permitted for weight (wt.), ounce (oz.), pound (lb.).

(f) The statement shall be in conspicuous and easily legible bold face print or type in distinct contrast (by typography, layout, color, embossing or molding) to other matter on the label. The letters may be no more than 3 times as high as they are wide.

(g) The statement shall include the words *"net weight"* or *"net wt."*

(h) If desired, (but not required), an *additional* weight statement in terms of the metric system may appear on the principal panel or elsewhere on the container.

(i) On a multi-unit retail package, a statement of the quantity of contents shall appear on the outside of the package and shall include the number of individual units, the quantity of each individual unit and in parenthesis, the total quantity of contents of the multi-unit package in terms of avoirdupois ounces.

5. Ingredient Statement — Not required if product is solely honey. If product is a mixture or blend of honey with any other substance or substances, the ingredients must be listed by common or usual names, in order of decreasing predominance. This ingredient statement may appear on any appropriate part of the label but the entire statement must appear on a single panel of the label.

6. Nutritional Claims — If you make any nutritional claims for your product *(but only if you make such claims)*, your label must carry a declaration of nutrition information under the heading "Nutrition Information Per Serving (Portion)." The terms "Per Serving (Portion)" are optional and may follow or be placed directly below the terms "Nutrition Information."

7. Illegal Representation on a Label.

(a) Labelling a *mixture* of honey and another food (e.g. corn syrup) and calling it "Honey" even though the other ingredient is listed in an ingredient statement.

(b) Any representation on the label that expresses or implies a geographical origin of the honey unless that representation is a truthful representation of geographical origin, is a trademark or a trade name which has so long and exclusively been used by a

> packer or distributor of that honey that it is generally understood by the consumer to mean the product of a particular packer or distributor or is so arbitrary or fanciful that it is not generally understood by the consumer to suggest geographic origin.

By a ruling of its Attorney General and Legislative Counsel, the State of California requires all containers of imported honey be labeled with the name of the country of origin even though repackaged. Containers of blended foreign and domestic honey must be labeled to state the product is such a blend and name the foreign country.

Adulteration of Honey*

The purity of honey is protected by federal pure food laws. Honey is a product the consumer knows and trusts — but it has not always been. Today, it is difficult to appreciate how important pure food laws were to beekeepers several decades ago. After glucose became common and cheap in the middle 1800's, adulteration occurred so frequently that the consumer was wary of purchasing extracted honey. Comb honey was the "real" thing and only it could be trusted as pure.

Although adulteration and substitution were common practice in the food industry, the honey producers were the first to feel the competition strongly enough to seek legal protection. In 1897 the United States Beekeepers' Union was formed for the purpose of combating the adulteration of honey, the defense of the beekeepers' legal rights, and the prosecution of dishonest commission men. Four states, Michigan, Minnesota, Kentucky, and New Jersey immediately passed laws prohibiting the adulteration of honey.

By 1906 the Federal government found that legislation in this area was imperative, and the Pure Food Act was passed by Congress; amended in 1938 as the Federal Food, Drug and Cosmetic Act, this law rendered the manufacture of the adulterated food or drugs in the United States, District of Columbia, and U.S. Territories unlawful. Following the bill's passage, public confidence in "pure" honey gradually returned and the market stabilized. Comb honey, enjoying its heyday before passage of the law, is now difficult to find in some markets.

---

*Renee Potosky and Dewey M. Caron, University of Maryland, Beekeepers and the Government, American Bee Journal, June 1977

CHAPTER XIII

DISEASE LEGALITIES

To set budding more
And still more later flowers for the bees
. . . . . John Keats

In the chapter Statutes, Ordinances and Regulations a compilation of interstate laws of the United States, on a state to state basis, was presented. This chapter will dwell on "bee disease laws" and court decisions pertaining to same.

Many statutes have been enacted to prevent or eradicate diseases of bees in the furtherance of the general welfare.[1] It is within the police power of a state to prescribe and enforce regulations to prevent or "stamp out" the spread of bee diseases for the publics' good. A Michigan court ruled that a:[2]

> "Statute designed to prevent spread of bee diseases by prohibiting bringing bees into the state on combs, used hives, or other apiary appliances in the interests of public health constitutes valid exercise of police power."

In a similar vein, in California:[3]

> "An act for prevention of bee diseases is proper exercise of police power although bee diseases are not harmful to humans, since it tends to promote the general welfare and is therefore a proper exercise of police power."

The following is compiled from Beekeepers and the Government (1977):[4] "There are state laws and regulations in nearly every state regarding bee diseases; there are no Federal regulations on diseases or their control. As of the states, their regulations vary from one to another. They attempt to regulate movement and entry of bees into the state, including issuance of permits and certificates and cover disease control treatments.

Thirty-eight states require that hives have movable frames. All but six states have apiary inspection regulations and beekeepers are compelled to permit the inspector right of entry. Only about half of the states require owner identification in an apiary. Sixteen states require state inspection of the honey house.

Bee colonies found infected with American foulbrood (AFB) are subject to different regulations and treatment in the various states. Thirty-seven states quarantine apiaries with AFB. The disease in colonies must not be concealed in fifteen states, must be declared in fourteen, cannot be exposed in twenty-six states, cannot be sold or transferred in forty states and must be destroyed in all but nine states (provided proper remedies are available to protect the rights of the individual apiarist). Although destruction of AFB colonies is required in most states, drugs are permitted for prevention of the disease in thirty-five states and for disease control in twenty states.

Bees that are moved are subject to regulation. Thirty-one states require a permit (usually issued only if apiaries are free of AFB disesase) for movement of bee colonies and bee equipment entering the state. Package bees and queens shipped interstate are subject to regulation as well.

Whereas the various states have laws and regulations to insure that disease is not moved from state to state, a federal statute helps insure that diseases from other countries are not admitted. Pollination services are not regulated except as it involves movement of bees and disease control.

Honey regulations are not strictly disease legalities but its close alliance merits some discussion.

The product honey is subject to few special regulations. Some states have regulations to limit the use of the word honey on "imitation" products made with isomerized corn syrup or with sugar blends using honey. The Pure Food Act, amended as the Pure Food, Drug and Cosmetic Act in 1938, rendered the manufacture of adulterated food or drugs in the United States, District of Columbia, and United States Territories unlawful. Following the bill's passage, public confidence in "pure" honey stabilized.

General laws of the food industry apply to packaging and labeling of honey. Honey grades and standards have been established by both the United States Department of Agriculture and the Food and Drug Administration."

Where a statute so provides it may be an offense to sell diseased bees. Usually it is held that the vendor's knowledge of the disease is an essential element,[5] but this element sufficiently exists if the vendor has knowledge of such fact as will put him on notice and cause knowledge of the condition to be imputed to him, as stated in a Missouri case:[6]

> "If reasonable care and caution would disclose the truth, the vendor cannot claim ignorance for he cannot recklessly shut his eyes to conditions and symptoms which if investigated would disclose that there was a diseased condition."

However, such statutes to be valid must not be unreasonable, arbitrary, or unjustly discriminatory,[7] and must not deny to the individual apiarist, affected thereby, due process of law, nor may they be violative of any other constitutional requirements or inhibition.[8]

CITATIONS

1) — *Carroll v. Tarburton,* Del. Supra., 209 A.86
2) — *Wyant v. Figy,* 66 N.W.2d.240, 340 Mich.602
3) — *Graham v. Kingswell,* 24 P.2d.488, 218 Cal.658
4) — *Renee Potosky and Dewey M. Caron,* American Bee Journal, June 1977
5) — *Wells v. Welch,* 224 S.W.120, 205 Mo.App.136
6) — *Ibid.*
7) — *Ex parte Goddard,* 190 P.916, 44 Nev.128
8) — *Id,* Del. Supra., 209 A.86

CHAPTER XIV

STATE DISEASE LAWS

The rules of the game
are what we call the laws of Nature.
. . . . . Thomas Henry Huxley

The first apiary inspection law in the United States was established in San Bernardino County, California in 1877. By 1883, a statewide law was passed by the California legislature, and by 1906, 12 states had laws relating to foulbrood. At present, most all States have laws regulating honey bees and beekeeping.

State laws and regulations relating to honey bees and beekeeping are designed primarily to control bee diseases. Therefore, they usually attempt to regulate movement and entry of bees, issuances of permits and certificates, apiary location control and quarantine, inspection, and methods of treating diseased colonies. These laws and regulations are summarized in the chapter Statutes, Ordinances and Regulations.

Most of the States require registration of apiaries, permits for movement of bees and equipment interstate,* certificates of inspection, right of entry of the inspector, movable-frame hives, quarantine of diseased apiaries, notification of the owner upon finding disease, prohibition of sale or transfer of diseased material, and use of penalties in the form of fines, jail or both. Although the destruction of American foulbrood diseased colonies is included in most state laws, these states also allow the use of drugs for control or preventative treatment of this disease.

The key figure in the enforcement of bee laws and regulations is the apiary inspector. He may have the entire state, a county, or a community under his jurisdiction. His duties and responsibilities are many and varied and upholding the state's regulations is his main objective.[2]

These duties of the apiary inspector, labeled State Apiarist in many States, are, in the majority of the States, quite similar. For a generalized understanding of regulations pertaining to the State Apiarist, statutes of the Commonwealth of Virginia are outlined below:[3]

> *Duties of State Apiarist:* The duties of the State Apiarist shall be to promote the science of beekeeping by education and other means, to inspect or cause to be inspected apiaries, beehives, and beekeeping equipment within the State for bee disease and to perform such other duties as may be required by regulation or law, including the inspection of honey houses for sanitation.

---

*Missouri was removed from the unenviable position of being the only state in the lower forty-eight without an effective bee-inspection law when Senate Bill No. 683 was signed by the Governor in 1978. Eighteen years of persistent efforts by members of the Missouri State Beekeepers' Association finally came to pass upon the signing of SB 683 (Missouri Apiculture Law).[1]

Inspections: measures to be taken for eradication or control of disease: The Commissioner or his assistants shall examine or inspect the bees in this State whenever they are suspected of being infected with American foulbrood disease or other contagious disease of bees, and shall inspect bees to be sold or to be transported interstate when requested.

If bees are found infected with American foulbrood or other contagious disease of bees, the State Apiarist shall cause suitable measures to be taken for the eradication or control of such disease. Whenever the owner of such diseased bees fails or refuses to take such steps as may be prescribed by the State Apiarist to eradicate or control the disease, the State Apiarist shall destroy or treat or cause such bees, together with the hives and honey, to be destroyed or treated in such a manner as he may deem best.

Appeal from order of State Apiarist: Any owner of diseased bees may, within ten (10) days from the receipt of an order from the State Apiarist to destroy or treat his diseased bees, hives or appliances, file a written appeal from the order with the Commissioner. The Commissioner, upon timely receipt of a written appeal under this section, shall act upon the appeal in accordance with the provisions of chapter 1.1. (Code of Virginia).

Right of entry for purpose of enforcement of chapter: The Commissioner, or his duly authorized assistants, shall have the authority, for the purpose of carrying out the provisions of this chapter, to enter upon, during reasonable business hours, any private or public premises, except private dwellings, and shall have access, ingress and egress to and from all apiaries or places where bees, combs and beekeeping equipment or appliances are kept, or where honey is being processed, stored or packed for market.

Notice of diseased bees: Any person in this State receiving knowledge of diseased bees in his or other apiaries shall immediately notify the State Apiarist or his assistant, giving the exact location where such diseased bees are kept, together with such other known information as may be requested.

Removal of bees, etc., from infected apiary may be forbidden: The Commissioner or his duly authorized assistants shall have authority to forbid the removal of bees, honey, wax, combs, hives or other used beekeeping equipment from any apiary or place where bees are known to be infected with American foulbrood or any other contagious disease, until he shall find the apiary free from disease and issue a certificate to that effect.

Rearing queen bees: No person in the State engaged in rearing queen bees for sale, shall use honey in the making of candy for use in mailing cages. Every person engaged in rearing queen bees shall have his queen rearing and queen mating apiary or apiaries inspected at least once during each summer season by the State Apiarist or inspector and on the discovery of the existence of any disease which is infectious or contagious in its nature and injurious to bees in their egg, larval, pupal or adult states, such persons shall at once cease to ship queen bees from such diseased apiary until the State Apiarist shall declare the apiary free from disease and issue a certificate to that effect.

Rearing queenright or queenless package bees: No person shall engage in the rearing of queenright or queenless package bees for sale without first applying to the State Apiarist for inspection at least once during each summer season. Upon the discovery of the existence of any bee diseases which are infectious or contagious and injurious to bees in any living stage, the rearer or seller shall at once cease to ship bees from affected apiaries until the State Apiarist shall declare the apiary or apiaries free from disease and issue a certificate to that effect.

Permit to bring bees and used bee equipment into State; inspection after bees enter State: No person shall bring into this State any bees on combs, empty used combs, used hives or other used apiary appliances from any other state, territories or country without first having received a permit so to do from the State Apiarist. This permit shall be issued only upon receipt of satisfactory proof that said bees and used beekeeping appliances are free from disease. Said permit shall be attached to the outside of the container of bees or goods so transported. Bees brought into this State shall be subject to inspection at any time after entering this State, at the discretion of the State Apiarist. Said permit shall have been issued on specifically identifiable colonies within sixty (60) days of shipment.

Resisting or hindering officials: It shall be unlawful for any person to resist, impede or hinder the Commissioner or his assistants in the discharge of their duties.

Certificate to accompany bill of sale: No bees on combs, hives, used beekeeping equipment with combs or appliances may be sold or offered for sale unless each bill of sale is accompanied by a valid certificate prepared by the Commissioner declaring the specifically identifiable hives, beekeeping equipment and appliances free of the causal agent of American foulbrood disease or other diseases named in the rules and regulations.

Abandoned apiaries: When an apiary is deemed by the State Apiarist to be abandoned, he shall certify his findings to the treasurer of the political subdivision in which the apiary is located, who shall give notice of such certification to the last known owner and the owner of the land upon which the apiary is located, by personal service, by posting at last known residence, or by publication. If, after sixty (60) days, the owner or landowner has not laid claim to said apiary, the treasurer may hold a sheriff's sale and issue a treasurer's deed to the successful bidder. The proceeds from such sale shall be added to the general fund of the political subdivision wherein the apiary is located. If such disposition of said apiary is not made within ninety (90) days of the date of the State Apiarist's declaration of abandonment he may take possession of the apiary and destroy the bees, hives and equipment therein.

Unsanitary conditions in operation of honey houses or buildings: Whenever it is determined by the State Apiarist or inspector, that unsanitary conditions exist in the operation of any honey houses or buildings or portion of a building in which honey is stored, graded or processed, the operator or owner of such honey houses or buildings shall be first notified and warned by the State Apiarist or inspector to place such honey house or building in a sanitary condition within a reasonable length of time. Failure to correct unsanitary condition after notification by the State Apiarist shall constitute a violation of this chapter and shall be dealt with as provided herein.

Violation of chapter: Any person violating any of the provisions of this chapter or any order or regulation issued by authority of this chapter or interfering in any way with the duly appointed representatives of the Commissioner of Agriculture and Commerce of this State in the discharge of the duties herein specified shall be deemed guilty of a misdemeanor and upon conviction shall be punished as provided in Section 18.1-9. (Code of Virginia).

In the State of Michigan a beekeeper commenced an action against the Michigan Department of Agriculture to restrain the enforcement of a statute regulating the manner by which bees may be imported into the state. He charged that the State of Michigan may not unreasonably exclude from importation articles of commerce . . . . . . He further stated — "the statute amounts to a state regulation of interstate commerce and involves a discrimination against interstate commerce."

The judges of the Michigan Supreme Court, in a 1954 decision, ruled:[4]

> "that the statute in question does not prevent the importation of bees and honey but does prevent their importation on combs, in used hives, or in used equipment. It is a general rule that a law enacted by the State to prevent the spread of contagious or infectious diseases among animals is a valid exercise of police power. Moreover, the right of properly constituted public authorities to destroy property deemed likely to spread disease is well established.
> In our opinion the State, in the exercise of its police power, has the right to regulate the manner by which bees may be imported into Michigan. The requirements under the act are in the interest of public health, and as such are within the constitutional provisions and did not constitute unconstitutional burden upon interstate commerce."

The effectiveness of bee laws and regulations is based on the compliance of the beekeeper. In the final analysis, responsibility for disease control remains with each individual bee raiser. Colonies are to be routinely examined for disease as a regular part of a good management program and the necessary steps taken whenever disease is found.

CITATIONS

1) — *C. R. Crain,* Missouri Passes Bee Law, American Bee Journal, August, 1978
2) — *A. S. Michael,* Bee Laws of the United States, American Bee Journal, July, 1976
3) — *Code of Virginia,* 1950 as amended, Statutes 3.1-610.3, 3.1-610.5, 3.1-610.6, 3.1-610.7, 3.1-610.8, 3.1-610.11, 3.1-610.12, 3.1-610.13, 3.1-610.15, 3.1-610.16, 3.1-610.17, 3.1-610.18, 3.1-610.19, 3.1-610.21
4) — *Wyant v. Figy,* 66 N.W.2d.240, 340 Mich.602

CHAPTER XV

IMPORTATION OF BEES

His helmet now shall make a hive for bees
. . . . George Peele

The importation of honeybees and honeybee semen into the United States from foreign countries is controlled by federal legislation. This importation of honeybees is prohibited except (Public Law 94-319):

(a) In order to prevent the introduction and spread of diseases and parasites harmful to honeybees, and the introduction of genetically undesirable germ plasma of honeybees, the importation into the United States of all honeybees is prohibited, except that honeybees may be imported into the United States:—

- (1) By the United States Department of Agriculture for experimental or scientific purposes, or
- (2) from countries determined by the Secretary of Agriculture —
  - (A) to be free of diseases or parasites harmful to honeybees, and undesirable species or subspecies of honeybees; and
  - (B) to have in operation precautions adequate to prevent the importation of honeybees from other countries, where harmful diseases or parasites, or undesirable species or subspecies of honeybees exist.

(b) Honeybee semen may be imported into the United States only from countries determined by the Secretary of Agriculture to be free of undesirable species or subspecies of honeybees, and which have in operation precautions adequate to prevent the importation of such undesirable honeybees and their semen.

(c) Honeybees and honeybee semen imported pursuant to subsections (a) and (b) of this section shall be imported under such rules and regulations as the Secretary of Agriculture and the Secretary of the Treasury shall prescribe.

(d) Except with respect to honeybees and honeybee semen imported pursuant to subsections (a) and (b) of this section, all honeybees or honeybee semen offered for import or intercepted entering the United States shall be destroyed or immediately exported.

Sec.3 (a) The Secretary of Agriculture either independently or in cooperation with States or political subdivisions thereof, farmers' associations, and similar organizations and individuals, is authorized to carry out operations or measures in the United States to eradicate, suppress, control, and to prevent or retard the spread of undesirable species and subspecies of honeybees.*

*Public Law 94-319, 94th. Congress, S.18, amend. June 25, 1976

The latest amendment also authorizes cooperative work with several other Central American countries for the eradication, suppression, control and prevention or retardation of the spread of undesirable species and subspecies of honeybees such as the Africanized honeybee in South America.

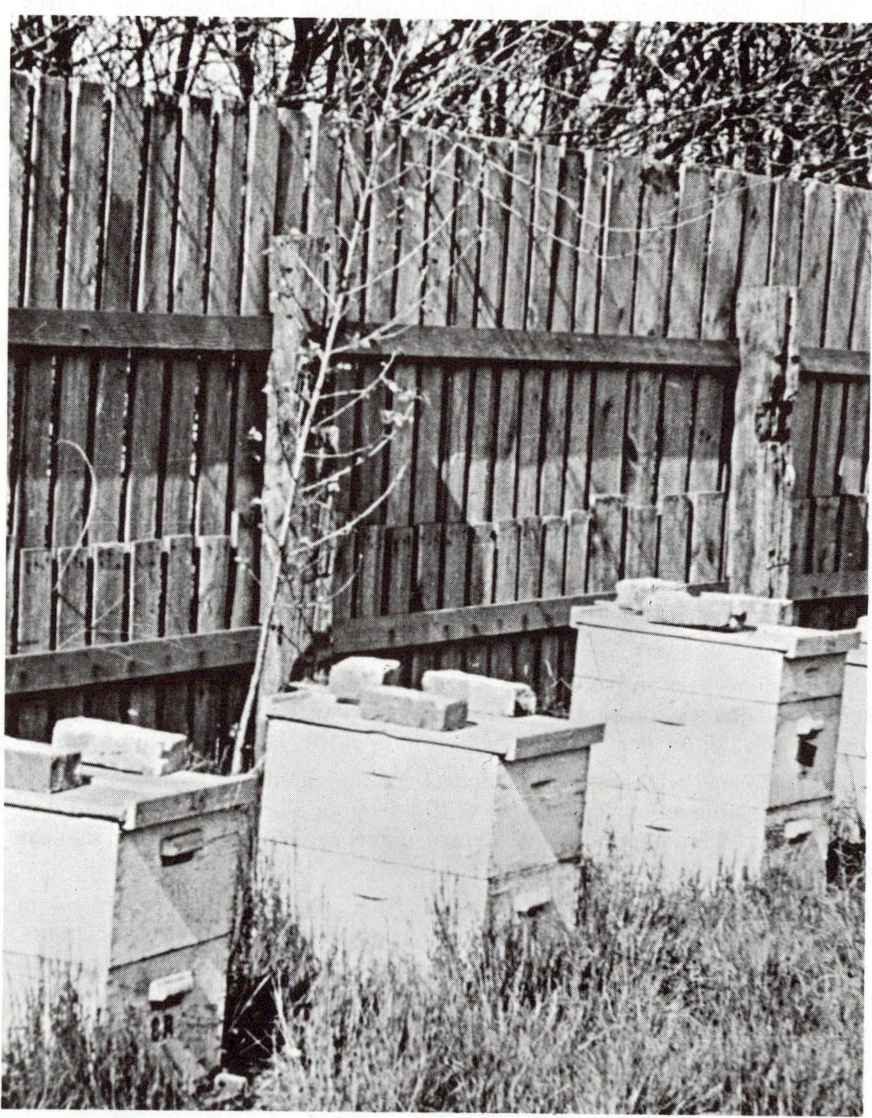

Fig. 8. The fence behind the hives will force the flight of the bees up and over neighborhood traffic, provide a winter windbreak, and will create a calm mini-environment which will help calm the bees.

CHAPTER XVI

GOVERNMENT PROGRAMS

We have rather chosen to fill our hives with honey and wax; thus furnishing mankind with two of the noblest of things, which are sweetness and light.

. . . . . Jonathan Swift

Crop Honey Loan and Purchase Program

A helpful federal program to the honey industry is the Crop Honey Loan and Purchase Program. Various bureaus of the United States Department of Agriculture offer information and assistance for the implantation of this program.

On a year to year basis, loans and purchase rates are announced by the Commodity Credit Corporation of the United States Department of Agriculture. Since the early '50's, though not annually, this program has been beneficial to producers and packers in establishing a floor price for honey.

The basic material provisions of the regulations containing the requirements with respect to loans and purchases under the Crop Honey Loan and Purchase Program are:

1. Loans: Producers must request a loan on the year's crop eligible honey usually on or before the 31st. of March.

2. Purchases: Producers desiring to offer eligible honey, not under loan for purchase, must complete a purchase agreement (CCC form) at the County Agricultural Stabilization and Conservation Service (ASCS) office usually on or before the 30th. of June of the following year. This agreement must specify the approximate quantity the producer desires to sell.

3. Unless demand is made earlier, loans on honey will mature on the 30th. of June of the following year.

4. The rate for the quantity of the year's crop honey placed under loan or acquired under loan or purchase shall be the rate for the respective class and color:

Table Honey	Cents per pound (1977)
a) white and lighter	33.5
b) extra light amber	32.5
c) light amber	31.5
d) other table honey	29 5
e) non-table honey	29.5

There is also a provision for the settlement value for a lot of honey which grades substandard on account of objectionable flavor, fermentation or carmelization or which grades substandard on account of defects or moisture or a combination of defects and moisture. The provision likewise sets forth who is an "eligible producer"; what is eligible honey; storage regulations; fees and charges; interest rates; and miscellaneous requirements.*

This program has contributed to a more stable honey market price over the past years and acts as a deterrent to drastic price declines.

Indemnity Payment Program

Many fruit, nut, vegetable, and seed crops require pollination by insects. The production of such crops is often dependent upon the availability of honeybees as primary pollinators.

Accidental destruction of bees and other pollinators have resulted from the use of insecticides and/or pesticides for the control of destructive insects. Because of the role of bees in crop production, their loss is not only potentially disastrous for the apiarist — it may also inhibit crop production in the area.

The object of the Beekeeper Indemnity Payment Program is to compensate beekeepers who, through no fault of their own, have suffered losses of honey bees as a result of the application of insecticides and/or pesticides. And, establish to the satisfaction of the county committee that they meet all the requirements for eligibility, such as:

(1) That during the application period, a loss of bees was suffered;

(2) That the loss of bees was caused solely by the use of insecticides and/or pesticides near or adjacent to their apiaries, and occurred without their fault;

(3) That if they used insecticides and/or pesticides, such use of chemicals in no way contributed to the loss of their bees;

(4) That if they had advance knowledge that insecticides and/or pesticides were to be used near or adjacent to their apiaries, they took reasonable precautions to protect ththr bees from exposure to the chemicals, or, if they took no such precautions, the failure to do so was reasonable under the circumstances, and;

(5) That after exposure of the bees to insecticides and/or pesticides they took reasonable action to minimize the bee loss to the extent that such action was feasible.

Beekeepers, to be eligible for an indemnity payment, shall, no later than July 15th. of each year, submit to the ASCS county office a signed statement specifying the number of colonies of bees and queen nuclei maintained at each apiary and the location of each apiary.

Beekeepers shall submit, within three (3) days, to the county committee an executed ASCS form, whenever there is a loss of bees as the result of

---

*Commodity Credit Corporation
United States Department of Agriculture
Federal Register — July 1, 1977

insecticides and/or pesticides, specifying the number of colonies destroyed, severely damaged, and moderately damaged; the number of queen nuclei destroyed; and the evidence of the loss of bees.

The county committee determines the amount of the indemnity payment(s) due to beekeepers whom it has determined to be in compliance with the terms and conditions of the program. The beekeeper's indemnity payment(s) is to be computed on the basis of the following rates of losses which have occurred since January 1, 1974:

(1) $22.50 for each colony destroyed
(2) $15.00 for each colony severely damaged
(3) $ 7.50 for each colony moderately damaged, and
(4) $ 7.50 for each queen nucleus destroyed.

Beekeepers must report losses immediately to their county ASCS offices so that appraisals may be undertaken. Failure to report losses within three (3) days of the damage may result in beekeepers being ruled ineligible for payments.

The Food and Agriculture Act of 1977 extended the Beekeeper Indemnity Payment Program to December 31, 1981.*

Since early 1979, there have been unfavorable rumblings in Congress concerning the Beekeepers Indemnity Payment Program. One Representative labeled it "a honey of a ripoff". James McIntire, director of the Office of Management and Budget, has reported that the said Beekeepers Indemnity Payment Program was one of several being eliminated.

---

*United States Department of Agriculture
Agricultural Stabilization and Conservation Service
39 F. R. 20063
Federal Register — June 10, 1974

Fig. 9. A properly equipped beekeeper inspects a healthy, orderly colony. Good management techniques are essential to produce a good honey crop — and help the beekeeper stay out of trouble.

CHAPTER XVII

TAXES AND THE BEEKEEPER

There was an Old Man in a tree,
Who was horribly bored by a bee;
When they said, "Does it buzz?"
He replied, "Yes, it does!
It's a regular brute of a bee!"
. . . . Edward Lear

Payment of taxes may be a three-pronged affair for the beekeeper — whether his apiary is operated as a commercial, side-line, or hobby venture. Taxes may be required to be paid to the federal government, state and municipality.

Whereas the federal government invariably stands first in line for collection of its taxes, the federal income tax will be discussed ahead of the others.

[1]The tax status of the professional beekeeper is obvious from his income tax return — beekeeping is his business. Not so for the side-line or hobby beekeeper. The problem these men face "Is this business entered into for profit or not?"

The Internal Revenue Service states most clearly that any net income (an excess of receipts over expenses) is taxable income, regardless of the source. In other words, keeping bees solely as a hobby and selling some of the surplus honey makes the beekeeper liable for taxes if the receipts from the sale of the honey is in excess of the expenses incurred in producing the honey.

On the other hand, if a hobbyist determines that the cost of producing honey was in excess of the receipts from sale, he CAN NOT deduct the "loss". Under the Internal Revenue Code, hobby losses are not allowable deductions. The philosophy behind this attitude is that hobby expenses are classed as personal living expenses (recreation) and personal living expenses are not allowed as a tax deduction. Hence, the hobbyist must pay income taxes on any profit but can not deduct any losses.

In order to qualify for that loss deduction, the beekeeper must switch his operation from hobby beekeeping to side-lining. In other words he must make a business out of it. The Internal Revenue Service does accept the fact that a person can be engaged in more than one trade or business at the same time.

But how big an operation is necessary so as to be classed as a business?

Short of a definite IRS ruling the answer to the question of size could be settled along the following lines: A beekeeper would be designated as pursuing a business for profit or livelihood if he—

(a) produces more honey than he could possibly consume, or reasonably be expected to give away to friends, relatives, neighbors, fellow workers, or dispose of in like fashion.

(b) who actively solicits outlets and customers for his honey; and,
(c) carries out his operations and sales in a businesslike manner.

[2]The beekeeper, now a honey producer, can have the option of computing his income and expenses on a cash or accrual basis. Once the choice is made for either the cash or accrual basis, the beekeeper must be consistent. Should he desire to change from one basis to the other, approval is to be secured from the Internal Revenue Service.

Generally speaking, the cash basis requires the least record keeping and is the simplest. Professional advice should be sought if the honey producer feels that some of his income falls in questionable areas. Consistency from year to year in the method of computing income and expenses carries weight with the IRS.

Under the accrual basis, income is recorded when it is earned regardless of whether the cash is actually received then, and expenses are recorded when they are incurred. Under this basis, one must keep detailed records of inventories and calculate the income on the cost of sales, beginning and ending inventories of honey which must be valued at market price less direct costs of disposition. The accrual method is somewhat more complex, but it will give a more representative picture of how honey production business is doing from year to year.

Some of the possible items of income for the honey producer are:

- Honey sales
- Beeswax sales
- Pollination fees
- Coop participation certificates when declared
- Interest
- Storage or rent
- Custom extracting
- Salaries or wages received

Expenses which the honey producer might incur:

- Package bees and queens
- Truck expenses - gas, oil, grease, repairs, and tires
- Freight
- Repairs on building and equipment
- Liability, fire, and vehicle insurance premiums - advance premiums should be pro-rated
- Heat, power, and water
- Telephone - portion pertaining to business
- Wages
- Property taxes - pertaining to business
- Small tools
- Sugar and drugs
- Miscellaneous shop supplies
- Travel expense - pertaining to business
- Office expense - related to business
- Auto expense - % deduction dependent upon its use for business or standard mileage deduction
- Container costs
- Rental equipment
- Fees and dues - related to the business

Beeyard rent investment
Interest expense
Periodicals - relating to beekeeping
Conventions and seminars - beekeeping related
Employer portion of Social Security Tax on employees

Depreciation expense may be taken on all fixed assets. The beekeeper is encouraged to set up a salvage value on the assets and depreciate the difference between the cost and the salvage value. The following is the suggested life of some fixed assets:

Buildings - dependent upon the type of structure erected	20-30 yrs.
Bee equipment	15 yrs.
Bees - commonly depreciated in areas where wintered	3 yrs.
Autos	3-5 yrs.
Trucks	5-10 yrs.
Shop equipment including extracting equipment	10 yrs.
Boilers	10 yrs.

The beekeeper has a choice of a number of methods of depreciation with straight line the most common. This method charges that portion of the asset as an expense each year, less salvage value divided by its life. Straight line and other methods are to be looked into, either in the material furnished by the Internal Revenue Service or by an accountant of your choice.

Income tax laws are complex and ever changing. The aforesaid information has touched on but a few of the items as they might affect the producer. Income tax problems vary with the operation of each individual beekeeper. It is good advice, if one has a specific income tax problem, to consult the nearest Internal Revenue office or one's personal accountant.

Many states have a state income tax imposed on their citizens and businesses. And this tax is likewise thrust on the honey producer. Other taxes are levied by some states on the beekeeper. A few examples are noted.

In the state of Wisconsin, an occupational tax is exacted against one or more colonies of bees - the sections state:

(1) There is imposed an annual occupational tax on every person, firm or corporation owning one or more colonies of bees of 25 cents for each colony in his possession or under his control. A colony of bees shall consist of a live queen or queen cell or cells, brood and adult bees, along with bottom board, cover, and one or more hive bodies with not less than 8 frames of comb. Bees and all bee equipment shall be exempt from all property taxes, but by *February 1* of each year the Department of Agriculture, *trade and consumer protection,* shall furnish to the state supervisor of assessments a list of counties and taxation districts of the owners of colonies as shown by the records of the Department.

(2) The occupational tax herein provided for shall be assessed to the owner or person in possession of such bees by the assessor. He shall enter on the assessment role the name of the person to whom assessed and the number of colonies. The clerk of taxation district shall compute the tax and enter it on the tax role. Such tax shall be collected in the same manner as taxes on personal property are collected. Twenty-five per cent of the tax shall be retained by the taxation district in which the bees are kept, and the balance shall be accounted for and paid to the state treasurer, in the same manner as state taxes on property are paid.

(3) At the request of the Department of Agriculture, *trade and consumer protection,* the clerk of the taxation district shall furnish the Department a list of the names and addresses of the beekeepers in his taxation district.

The State of Idaho lists an assessment tax - its sections follow:

(a) There is hereby levied and imposed upon each colony or hive of bees within the State of Idaho on July 1st. of each year a continuing annual tax of three cents per hive or colony of bees beginning in the year 1970 for the purpose of carrying on the provisions of chapter 25 of title 22, Idaho Code.

(b) The tax may be increased to not more than ten cents (10c) per hive or colony per year, if approved by a majority of beekeepers voting in a referendum held for the purpose of determining whether such levy of the tax shall or shall not be changed. If the levy or the tax is changed, the levy of the tax will continue annually at the changed rate until again changed by another referendum. Any resident of Idaho who is registered under this chapter as an Idaho beekeeper with the Idaho Department of Agriculture may vote at such referendum. Any referendum to be held for the purpose of changing the levy or such tax shall be held at the annual meeting of the Idaho honey industry association or any successor organization to this group.

(c) On or before the 15th. day of August of each year any person engaging in the business of apiculture shall make and file in writing with the Idaho Department of Agriculture a statement specifying the name, residence, place of business of the beekeeper, number of hives or colonies of bees owned or controlled, number of apiaries maintained and the location of such hives, colonies and apiaries by governmental subdivision or such other designation of location as may be provided for by the Department of Agriculture, and such other information as may be requested.

(d) The tax provided for in this section shall be due and payable on or before October 31st. of each year, and it shall be collected by the Idaho Department of Agriculture. Upon receipt of the annual statements provided for above in this section, the Idaho Department of Agriculture shall bill each beekeeper for said tax and it shall take the necessary steps to collect such tax, including civil action in proper courts.

(e) Said tax shall be a lien upon all apicultural products, equipment, bees and property of the person owning or controlling such bees and shall be prior to all other liens or encumbrances except liens which are declared prior by operation of the statutes of this state.

(f) The Department of Agriculture shall devise a system of identification for apiaries including a permanent number to be assigned to each beekeeper. The system may include placards which shall be permanently posted and maintained by each beekeeper at each apiary. The Department may also require that each hive be stenciled with the beekeeper's permanent number. If it does so, it shall issue stencils for that purpose.

A problem arises in many localities as to the "situs of property" for taxation purposes. In the State of Minnesota, their tax sections state:

"Unless otherwise provided by law, taxable personalty (includes hives) is listed and assessed in district where owner, agent or trustee resides. In case of doubt as to proper place of listing and assessment, or where listing cannot be made as provided by law, County Board of Equalization decides as between places in same county, while Commissioner has jurisdiction as between counties or places in different counties."

Disputes often occur — where is the situs of the property to be taxed? Two rulings in Minnesota answered:

> "Hive of bees kept in Red Lake County but owned by Carver County resident engaged in business of marketing and producing honey in Carver County should be listed and assessed in taxing district where owner resides."

And:

> "Because state or owner's domicile has right to tax his personalty which hasn't acquired fixed situs outside state, bees owned by resident on May 1st. may be taxed in Minnesota even though on that date they (bees) are temporarily outside state."

Some states accord the "farmer" status to beekeepers, other states do not. The Commonwealth of Virginia classifies the bee raiser in the same category as any other taxpayer — no farmer status. In this Commonwealth, the beekeeper, filing his state income tax form, will file as any other taxpayer. Thus, any financial benefits that may be received by an agriculturist, in the payment of state income taxes, is a benefit(s) not shared by Virginia honey producers.

Questions do arise on tax problems. Inquires to these questions should be addressed to the proper state authorities or, as most taxpayers do, acquire the services of a tax consultant or an accountant. These individuals are professionals in their field and their services may save money and grief for the apiary owner.

Taxes are also paid to the municipality. The county, city, or town imposes assesments on personal property, real estate, vehicles, etc. The honey producer, be he side-liner, hobbyist, or professional, pays taxes on his apiary equipment in some states; pays on his real estate, if he is the fee simple owner of land and buildings; pays a sticker fee (tax) if he owns vehicles; pays a business tax, if he sells his first pound of honey; and pays other taxes the municipality may levy.

A Nebraska Supreme Court case is presented because it affects the majority of beekeepers where it is most important — their pocketbooks. The proceeding involved the assessment of a personal property tax on certain colonies of bees. The decision was rendered in 1976.[3]

> "A partnership operates honey farms for the production of honey and beeswax in the States of Nebraska and California. In the fall of 1971, at the end of the season in Nebraska, in accordance with its customary practice, the partnership retained 250 colonies of bees in Nebraska for the winter. They transported approximately 800 colonies to California. The partnership reported the 250 colonies of bees for assessment in Sherman County, Nebraska, on the tax day of January 1, 1972. On March 1, 1972, the partnership reported the 800 colonies of bees for assessment in the State of California.
>
> Between January 1 and July 1, 1972, the number of retained colonies of bees in Nebraska increased by 750 additional through a process of culling, dividing and introducing new queens.
>
> Meanwhile in California, the 800 colonies were divided to make 4,800 colonies which were brought to Nebraska. Of these, 1,365 were brought into Sherman County. When the assessor learned of the additional number of colonies above those reported on January 1, the additional colonies were added to the partnership's assessment as omitted property.

The partnership protested this added assessment. At the trial hearing, Sherman County contends that under Nebraska statutes, where living creatures have been taxed as property in another state for the tax year involved and produce offspring, progeny or descendant in that state which are brought into Nebraska before July 1, such increase must be added to the personal property assessment rolls in Nebraska and taxed accordingly.

The administrator of the property tax division of the Department of Revenue of the State of Nebraska stated that it was the custom and the practice in Nebraska that no livestock born within Nebraska after the 1st. day of January is subject to assessment for tax purposes for that year. He also declared that livestock born after January 1 and brought into Nebraska before July 1 are not assessed. And further, in his opinion, bees brought into Nebraska should be treated the same for assessment and taxation purposes as livestock.

This complicated tax problem was tackled by the justices of the Nebraska Supreme Court — their decision:

"In substance, the evidence is uncontradicted that at least when the progenitors of living creatures have been assessed for personal property taxes in Nebraska or any other state for a given tax year, their progeny or offspring, born, hatched or brought into existence after the tax year, have not been in the past, and are not now, assessed for personal property taxation by any county in the State of Nebraska.

In the case before us, the evidence is clear that the assessment against the colonies of honey bees involved here was arbitrary and discriminatory, and that Sherman County's interpretation of the statute was not enforced against other taxpayers owning property of the same type, class, and character.

We find no reasonable distinction for tangible personal property tax purposes between livestock and honey bees or other living creatures owned and used for commercial purposes. The statutes of Nebraska make it clear that the legislature intended that personal property, properly assessed and held for taxes for that year in another state, and moved or brought into Nebraska on or before July 1, should not be assessed again for taxes for that year. Colonies of honey bees which were not in existence on January 1st., which are brought into Nebraska from another state before July 1st., are not subject to personal property tax assessment in Nebraska where their progenitors were listed and assessed for taxation for that year in another state."

CITATIONS

1) — *Robert J. Hehre,* C.P.A., Income Taxes and the Beekeeper, American Bee Journal, March 1970

2) — *Harry Rodenburg,* The Honey Producer and His Income Tax, American Bee Journal, March 1968

3) — *Knoefler Honey Farms v. County of Sherman,* 243 N.W. 2d.760, 196 Neb.435

CHAPTER XVIII

CONTRACTS

A swarm of bees in July
Is not worth a fly
. . . *Anonymous*

The law of contracts is complex, voluminous and variegated and no attempt will be made to cover the subject, if it ever can be covered, in a chapter or even in volumes. Informative highlights for the beekeeper will be touched upon in this chapter.

With all our legal knowledge no entirely satisfactory definiton of contract has been devised. A leading authority on contracts[1] has stated: "A contract is a promise, or set of promises, for breach of which the law gives a remedy or the performance of which the law in someway recognizes as a duty". Another common definition — "it is a legally enforceable agreement". And in a Federal proceeding, a contract was ruled to be "the coming together of two minds on a thing done or to be done".[2]

The bible of contractual law is The Uniform Commercial Code (UCC). It will be incorporated in this chapter for clarification on certain points of contractual law.

A contract is executory in nature. It contains a promise or promises that must be executed, that is, performed. For example — an agreement to sell a quantity of bees is a contract; the sale of this quantity of bees is not.

In the formation of a contract the basic requisites are:[3]

1. A promise by a party having legal capacity to act.
2. Two or more contracting parties — no man can contract with himself.
3. Mutual assent.
4. Consideration.
5. The agreement must not be void.

A noted Florida attorney[4] has written that "the subject of written contracts is an important one to beekeepers and deserves their renewed attention. It is not necessary that a written contract . . . be in any special form. But it is important that certain items be included in the written document for it to be a legal and binding contract. Points to be included (buy and sell):

(a) the document must be dated
(b) the document must specifically describe the items sold and purchased
(c) the purchase price must be stated
(d) the document should specifically state how payment is to be made (describing exactly where, when, at what rate of interest payments are to be made, if time payments are involved)
(e) if delivery of the goods is part of the agreement, it should give the details of delivery (who pays for delivery, who handles delivery, when and where the goods are to be delivered)

(f) the document should state what is to happen or what penalties are to be paid if either side later fails to honor any of the conditions of the contract
(g) the document must contain the signatures of all the parties to the contract (everyone who is required to do anything or pay anything under the terms of the contract)

Sometimes letters exchanged between two people may constitute a binding contract where the two letters together constitute an enforceable written contract."

The first requisite of a contract is that the parties manifest to each other their mutual assent to the same bargain at the same time. Their mutual assent will normally take the form of an offer and an acceptance. A Delaware court called attention to offer and acceptance, stating:[5]

> "To constitute a contract, there must be a offer made by one person to another and an acceptance of that offer by the person to whom it was made."

An offer, however, has to be distinguished from an expression of an opinion or a prediction, as ruled in a well-known case:[6]

> "A tenant farmer was behind in his rent. He told his landlord about his problems and the landlord suggested he should get more animals. The tenant replied, "If I stock up too heavy in the pasture and there be a short spell I will be up against it and that is the reason I am waiting for you."
> The landlord told his tenant, "Never mind the water John, I will see that there will be plenty of water because it never failed in Minnesota yet." On the strength of the landlord's statement, the tenant purchased 107 animals and lost money when the water supply failed.
> The court held, as a matter of law, that the landlord's statement was not a promise. It was nothing more than an opinion or a weather prediction. If the landlord had said, "I will make good any loss you suffer due to a water shortage," the result could well have been different even though the rainfall was not within the control of the promisor."

Let's look at another example of offer and acceptance. (A) wires, "I will not sell for less than $5000," and (B) wired, "Will accept your offer $5000 net." It was held that no contract resulted, since (A's) wire did not constitute an offer. Why?

(A's) wire may have meant that he is holding the land for sale at a minimum price of $5,000 hoping to obtain a higher price. He had not said, either that he will sell at that price or that he will sell at all. Nor will this be implied. Having made no promise, (A) has made no offer. It stands to reason that (B's) acceptance is valueless.

The eminent and learned Judge Cardoza identified three elements which must concur before a promise is supported by consideration:[7]

1. The promisee must do or promise to do what he is not legally obligated to do; or refrain from doing or promise to refrain from doing what he is legally privileged to do.
2. The promisor must have made the promise because he wishes to exchange it at least in part for the detriment to be suffered by the promise.
3. The promisee must know of the offer and intend to accept.

As there may be the making of a contract so there may be the breaking of a contract or, as is known in the law, a breach. It has been defined as the "failure, without legal excuse, to perform any promise which forms the whole or part of a contract".[8] In a contractual sale of bees in Minnesota, fraud was shown as the element of breach:[9]

> "Plaintiff represented that the bees "were all clean and free from disease", but that in fact they were afflicted with a disease known as "foul-brood" . . .
>
> Assuming the evidence sufficent to prove that the bees were in fact diseased, there is ample proof of all the necessary elements of fraud."

The general rule of law of contracts is that anything so material as to defeat the essential purpose of the parties is a breach. A buyer's refusal to accept or to pay in accordance with the terms of the contract is as clearly a breach as is the seller's failure or refusal to deliver goods or chattel conforming to the contract.[10]

A California proceeding brought out a facet of a contractual breach when the judges ruled:[11]

> "If bee raiser was first to breach the contract, then he is liable in this action for damages even if, after such breach by him, bee owner did take the automobile to his own home and assume the sole care of the bees. The fact that the bee owner removed the automobile from the premises that bee raiser had abandoned, and placed it where it would not be subject to the destructive effects of the sun's rays, and the further fact that bee owner cared for the bees that the bee raiser seemingly abandoned . . .
>
> And if, as might be inferred from his conduct in leaving the house and the automobile, bee raiser was under no obligation to demand the bee owner continue to perform his contract. Bee raiser cannot escape liability because bee owner did nothing other than to take steps to prevent a further loss and an increase in damages. The rule is that a party to a contract cannot take advantage of his own omission to observe the contract he cannot interpose the breach as a defense to an action on the contract . . .
>
> The rule in general, that the right to rescind a contract rests only with the party who is without default. One party cannot . . . violate the contract himself, and then seek a rescission on the ground that the other party has followed his example . . . A party to a contract first in fault cannot rescind it."

Contracts, though considered legal by the contracting parties, may be considered illegal for either one of two reasons: (1) because it is made illegal by statute, or (2) because it is against public policy as declared by the courts. The following statutes are the more important ones rendering illegality to a contract:

(a) statutes regulating the conduct of a particular trade, business or profession
(b) statutes regulating traffic in intoxicating liquors
(c) statutes prohibiting labor or business on Sundays
(d) statutes prohibiting the taking of usury
(e) statutes prohibiting gaming and wagers
(f) statutes prohibiting lotteries[12]

A lengthy Louisiana proceeding involved, among others, bees and a contract:[13]

> "A contract stated the following:
>
> > I, J. Lloyd St. Romain of Moreauville, Louisiana, sell 350 colonies of bees, 800 mating hives and saw to F. M. Morgan of Hamburg, Louisiana, with the agreement that the bees will be delivered November 1, 1947 or will pay the sum of $1,982.88 for the consideration of $1,700.00; of which I have received today November 1, 1945, $1,000.00 and balance to be given to me later in the regular course of business. (The $282.88 in excess of the principal sum of $1,700.00 represented interest.)
>
> In the latter part of 1945, St. Romain acquired a patron, Jean Louis Bechard, who was engaged in the business of beekeeping. The first business dealing between the two occurred in December 1945. Bechard mailed St. Romain a bank draft for $504.50 to pay for bees to be delivered on a future date. Thereafter, bees of the value of $54.50 were delivered. In spite of the default, Bechard, during February and March 1947, made advances aggregating $2,083.25, to pay the price of bees to be delivered on future dates.
>
> Not a bee was delivered for the advanced $2,083.25.
>
> After the default and amicable demand, a suit was filed to recover the total sum due Bechard ($2,083.25). A writ of attachment was placed upon the bee colonies and bees named in the agreement between St. Romain and Morgan. Under this writ 200 hives of bees were seized by the sheriff. St. Romain, in his answer, admitted owing the amount for which Bechard sued, but denies that any cause or reason existed for issuance of the writ of attachment. After all, the document signed by Morgan and himself was a bona fide contract.
>
> Morgan intervened in the suit and asserted ownership of the seized bees under the signed agreement. He professed it was part of the contract between St. Romain and himself that the bees were to remain in St. Romain's possession and under his supervision until the 1st. of November, 1947, at which time "Actual delivery would be made and possession given" to him.
>
> The allegations of Morgan were answered by Bechard. He attacked the purported act of sale from St. Romain to Morgan as being a simulation outright. That as an instrument of sale it was not effective as to him because the seller retained possession of the bees.
>
> Hearing the arguments and counterarguments, the Honorable Judges rendered their verdict:
>
> One of the indispensable essentials to the legality of an instrument of sale, with a right of redemption is that actual, physical possession of property involved pass to purchaser. It is obvious that use of the word 'sell' in said instrument was improvident. To have clearly expressed the intent of the parties, the word 'mortgage' should have been employed. The construction, as placed by the parties themselves upon the instrument over a period of nearly two years, unquestionably proves that the instrument was really designed to operate as a contract of security to protect Morgan for the amounts he advanced.
>
> Our conclusion, therefore, is that title to the bees did not pass to Morgan by virtue of the instrument herein discussed, and this being true, the bees were at all times the property of St. Romain and were liable to seizure for his debts.

The instrument not being of record, of course, was ineffective as to a seizing creditor and other third persons."

The court purported to have followed the old axiom — possession is nine-tenths of the law.

CITATIONS

1) — *Williston, Contracts,* Section 1, 3rd. Edition
2) — *Charles R. Sheperd, Inc. v. Clement Bros. Co.,* D.C. N.C., 177 F.S.288
3) — *Williston, Contracts,* Section 18, 3rd. Edition
4) — *Nelson E. Bailey,* Contract Law for Beekeepers, American Bee Journal, July 1976
5) — *Salisbury v. Credit Service,* Del.supra., 199 A.674
6) — *Anderson v. Backlund,* 199 N.W.90, 159 Minn.,423
7) — *Allegheny College v. Nat. Chautauqua County Bank,* 159 N.E.173, 246 N.Y.369
8) — *Friedman v. Kratzner,* 114 A. 884, 139 Md.195
9) — *Sampson v. Penney,* 197 N.W. 135, 151 Minn.411
10) — *Franklin v. Pence,* 36 S.E. 2d. 505, 128 W.Va.353
11) — *Ross v. Tabor,* Calif. App., 200 P.971
12) — *Law of Contracts,* 2nd Edition - Laurence P. Simpson, West Publishing Co., St. Paul, Minn.
13) — *Bechard v. St. Romain,* La. App., 25 So. 2d.606

Fig. 10. This colony is about to swarm. Although swarming is a natural instinct, it will take quick and immediate action on the part of the beekeeper to get his colony back into production and to avoid any type of incident that could create a lot of bad publicity.

CHAPTER XIX

LEGAL FORMS

Of what use are forms, seeing at times they are empty?
Of the same use as barrels, which, at times, are empty too.

. . . . . . *Hare*

The forms in this chapter are intended as mere guide-lines and no guarantee is given that they are all inclusive. Each particular problem, matter, agreement, contract, lease, etc. presents its own idiosyncrasy and an attorney should be consulted if there is any question whatsoever.

(A)

*BEE COLONY OWNERSHIP AGREEMENT

TO WHOM IT MAY CONCERN:

This document will serve to confirm that ________________ (Property Owner) of ________________________ (Address) and ________________ (Apiary Owner) of ________________________ (Address) have mutually agreed to allow ________________ (Apiary Owner) to locate an apiary on the property of ________________ (Property Owner).

Be it also established hereby that the bee keeping artifacts, hives, contents, etc. (list items) do not compose part of the property as permanent fixtures.

In the event of sale or rent (lease) of such lands (property) whereupon the apiary is situated, the said bee keeping artifacts, hives, contents, etc. (list items) do not become part of the lands (property) for sale or rent; but shall remain the property of ____________________ (Apiary Owner).

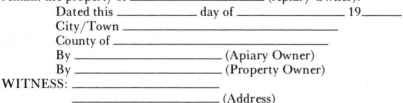

Dated this ____________ day of ________________ 19______
City/Town ________________________________
County of ________________________________
By ________________________ (Apiary Owner)
By ________________________ (Property Owner)
WITNESS: ________________________
________________________ (Address)

---

*John Miller, Ontario, Canada, American Bee Journal, March 1975

(B)

***CONTRACT TO KEEP AND MAINTAIN BEES.**

In consideration of ______(A)______ delivering possession of ______ hundred swarms of bees, ______ truck, honey tank and equipment for extracting honey, to ______(B)______, the receipt of which is hereby acknowledged, ____(B)____ agrees to keep and maintain said ______ hundred swarms of bees in a good condition and in such a state of condition as is usually practiced by a reliable bee man in the bee business, and that he shall not abandon or desert bees or any part thereof.

________(B)________ shall keep ______ truck and equipment for extracting honey in good repair at his own expense, and furnish all work and labor at his own expense and cost in connection with the care, upkeep, and maintenance of said bees and the extraction of honey, and within seasonable and reasonable time, he shall deliver to __________(A)________, ______ percent of all honey obtained from bees together with ______ percent of the wax, delivery to be made to warehouse most convenient to ______(B)______. ______(A)______ to furnish cans for his portion of honey.

At end of ____ years from date hereof, said ______ hundred swarms of bees, ______ truck, honey tank and equipment for extracting honey, shall become the property of ______(B)______, provided all conditions of this agreement have been fulfilled and complied with by ________(B)________, and provided further that amount of honey and wax ________(A)________ shall have received from ________(B)________ shall have at least amounted to market value of ______ dollars in excess of cost of cans and cases furnished by ____(A)____ for his portion of the honey, and provided further increase in and from the ________ hundred swarms of bees shall be equally divided between the parties hereto at end of ________ years.

In event that ____(B)____ shall become dissatisfied with this agreement, he shall return said __________ hundred swarms of bees together with ________ truck, honey tank, and equipment for extracting honey to ______________, the original location of said __________ hundred swarms of bees.

________________ (A) ________________ (SEAL)
(signature)

________________ (B) ________________ (SEAL)
(signature)

---

*Adopted from Gray v. Craig, 127 Cal. App. 374, 16 P. 2d. 798

(C)

*BILL OF SALE WITH AFFIDAVIT OF TITLE

KNOW ALL MEN that I _________________________ (seller's name) of ___________________ (City), in the County of ________________, State of ________________, in consideration of the sum of $________________ (purchase price), to me in hand paid by _________________________ (buyer's name) of _________________________ (buyer's address), the receipt whereof is hereby acknowledged, have granted, bargained, sold, transferred, conveyed, and delivered, and do bargain, sell, grant, convey, transfer and deliver unto the aforesaid ___________________ (buyer's name) the following goods and chattels, to-wit:

(Here describe the items sold; by type or use, identification marks and serial numbers, brand names, seller's brands or markings, and amounts or numbers of each item sold. If possible, the description should be clear enough that anyone could look at particular pieces of equipment or groups of beehives and then determine by reading the bill of sale whether or not these particular items were the ones described.)

to have and to hold the same unto ______________________ (buyer's name), his executors, administrators, and assigns forever;

And I do for myself, my executors, administrators, and assigns covenant and agree to and with ______________________ (buyer's name) to warrant and defend title to the goods and chattels hereby sold to ______________________ (buyer's name), his executors, administrators, and assigns against all and every person or persons whomsoever.

In witness whereof I have set my hand and seal this ______________ (date)

Signed, sealed and delivered this ______________ (date)

______________________________ (seller's signature) (L.S.)

STATE OF ______________________________)
) ss:
COUNTY OF ____________________________)

______________________________ (seller's name), being duly sworn, deposes and says:

1. That he resides at ______________________________ (seller's address), and is the sole owner of the goods and chattels described in the BILL OF SALE to which this affidavit is attached, and that he is the same person who executed the BILL OF SALE to which this affidavit is attached.

---

*Nelson E. Bailey, Buying Used Hives or Equipment? Get a Bill of Sale! American Bee Journal, February 1976

2. That he has full right to sell and transfer the goods and chattels described in the BILL OF SALE to which this affidavit is attached.

3. That such goods and chattels are free and clear of all debts, liabilities, obligations and incumbrances of any form whatsoever.

4. That there are no judgments against him in any court of the State of ______________, or of the United States of America, and there are no replevins, attachments, executions, or other process or writs issued against him in any form whatsoever; and that he has not filed any petition for bankruptcy, not has any petition for bankruptcy been filed against him and that he has not been adjudicated a bankrupt.

5. That this affidavit is made to induce ______________ (buyer's name) to accept the transfer of the goods and chattels described in the BILL OF SALE to which this affidavit is attached and to induce ______________ (buyer's name) to pay the consideration therefore as specified in the BILL OF SALE to which this affidavit is attached.

Witness my hand and seal in the county and state aforesaid on this ____________ day of ________________ (month), 19______.

____
NOTARY PUBLIC

My commission expires ______________________________

BILL OF SALE WITH WARRANTY

Know all men by these presents that I, ________________ of ________________, in the County of ______________, State of ______________, in consideration of ____________ ($) Dollars to me paid by ________________ of ________________, the receipt whereof is hereby acknowledged, hereby sell to ________________, ______________ (No., of swarms of bees, description of equipment).

To have and to hold the same to ____________, his executors, administrators, and assigns, forever. And I warrant the swarms of bees to be free of disease.

And I, for myself, my heirs, executors, and administrators agree with ________________ to warrant and defend swarms of bees sold to ________________, his executors, administrators, and assigns, against all and every person and persons whomsoever.

In witness whereof I have this ____________ day of ______________, 19______, affixed my signature and seal.

______________________________(SEAL)

*BEEKEEPER - GROWER CONTRACT OF RENTAL OF BEES FOR POLLINATION

To avoid any misunderstanding, the beekeeper and grower should enter into a written agreement on the services to be performed and the responsibilities of each party. Whereas many different and varied factors may enter into this type of agreement, experience had indicated that the following considerations should be incorporated in such a contract.

1. The number and strength of colonies that are to be provided.
   (a) Strength of colony should be indicated as frames of bees and square inches of brood adequate for the time of year.
   (b) The colonies must be free of disease and queen-right.
   (c) The colonies should be in standard hives.

2. The time at which the colonies are to be moved in, and the duration of the service.

3. The distribution of the hives in the fields or orchards, and the services to be performed by the beekeeper and farmer locating them. (It should be stipulated that the grower will give the beekeeper at least 72 hours notice of the time that the hives are to be moved in or out, since this will vary with the period of bloom.)

4. The responsibility of the farmer for loss of bees during the service from mechanical disturbance of the hives, from irrigation, or from chemical poisoning due to the application of pesticides.

5. In case pesticides which are toxic to bees have to be applied to the crops in bloom, notification of the beekeeper in time sufficient to allow him to remove the hives before the applications are made. An understanding should also be incorporated concerning the payment of the cost of moving the hives back again after the danger period from pesticides has past. (Usually the beekeeper moves them out and the grower pays the cost of having them moved in again.)
   (a) Since both the grower and the beekeeper are interested in the welfare of the colonies, both will want the bees protected from heavy loss due to the application of toxic substances on neighboring property on which the bees may work.

6. The rental price with terms of payment.

7. If the rental price is on a share basis, agreement by the grower to use approved methods of production and by the beekeeper to use proper manipulations of the colonies so as to maintain adequate populations and suitable working conditions for the bees.

(SIGNATURES)
____

*From BEEKEEPING by Eckert and Shaw
The Macmillan Company, New York, New York

The hazards of renting bees for pollination should be taken into consideration in the price structure quoted in the contract. These may be enumerated as follows:

1. There is about 5 per cent loss of queen through moving.

2. Occasional colonies are lost by overheating in transit.

3. Drifting of bees to other hives or from the center of the fields to hives nearer the edges is rather common.

4. Occasionally a wreck results in the loss of a major number of the hives and colonies being hauled.

5. Loss due to mechanical injury from farm equipment is not uncommon.

6. The heavy concentration of a large number of hives of different beekeepers results in overstocking a territory and in creating conditions favorable to robbing amongst the bees; this, in turn, is one of the main causes of the spread of brood diseases.

7. It is difficult to move a heavy concentration of hives quickly enough to avoid injury from pesticides applied in the area.

8. When bees are scattered in the fields or orchards, they cannot be given the efficient attention they would receive in permanent locations. This results in losses due to swarming, queenlessness, and lack of supers.

9. Overcrowding may necessitate the feeding of winter stores. (This seldom occurs in orchard colonies.)

10. Losses involving a majority of a beekeeper's colonies may seriously affect his income for the balance of the year and the following year.

(D)

ACKNOWLEDGMENT

STATE OF ____________________
COUNTY/CITY OF ____________________ to-wit:

I, the undersigned, a Notary Public in and for the County/City and State aforesaid, do hereby certify that ________________, whose name is signed to the foregoing writing, bearing date on the ____________ day of ________________, 19______ has acknowledged the same before me in my County/City and State aforesaid.

Given under my hand this ____________ day of ________________, 19______.

My commission expires ______________

_________________________ NOTARY PUBLIC

HUSBAND AND WIFE ACKNOWLEDGMENT

STATE OF ______________

COUNTY/CITY OF ______________ to-wit:

I, the undersigned, a Notary Public in and for the County/City and State aforesaid, do hereby certify that ___________ and ___________, husband and wife, whose names are signed to the foregoing writing, bearing date on the ___________ day of ______________ 19_____, have acknowledged the same before me in my County/City and State aforesaid.

Given under my hand this ___________ day of ______________ 19_____.

My commission expires ______________

______________________ NOTARY PUBLIC

CORPORATE ACKNOWLEDGMENT

STATE OF ______________

COUNTY/CITY OF ______________ to-wit:

I, the undersigned, a Notary Public in and for the County/City and State aforesaid, do hereby certify that ___________ and ___________, President and Secretary respectively of the ______________ Corporation, whose names are signed to the foregoing writing, bearing date on the ___________ day of ______________, 19_____, have acknowledged the same before me in my County/City and State aforesaid.

Given under my hand this ___________ day of ______________, 19_____.

My commission expires ______________

______________________ NOTARY PUBLIC

*POLLINATION AGREEMENT

For Season 19______

Date ______________

The Beekeeper
Name ______________________________
Address ______________________________
____

Phone Number ______________________

The Grower
Name ______________________________
Address ______________________________
____

Phone Number ______________________
Number of Colonies Ordered ______________________
Rental Fee for Grade A Colonies ______________________
Rental Fee for Grade B Colonies ______________________
Compensation for Additional Movement
of Bees or Other Extras ______________________
Total Rental Fee ______________________
Name of Crop ______________________
Location of Crop ______________________
____

Distribution Pattern of Colonies Shall Be ______________________
____

The Grower Agrees:

1. To give ________ days notice to bring colonies into the crop.
2. To give ________ days notice to take colonies out of the crop.
3. To pay one-half the agreed total fee when the bees are delivered.
4. To pay in full within ________ days after the delivery date.
5. To pay one (1) percentum a month interest on amounts unpaid after the due date.
6. To use no toxic pesticides (of any nature whatsoever) in the crop during the rental period except with the understanding and consent of the beekeeper, and to warn the beekeeper if neighbors use toxic pesticides.
7. To provide an uncontaminated water supply.
8. To assume liability for livestock damage or vandalism.
9. To assume public liability for stinging while the bees are on location in the crop.

***E. C. Martin,** The Use of Bees For Crop Pollination, The Hive and the Honey Bee, Dadant & Sons, Publishers.

The Beekeeper Agrees:

1. To open and demonstrate the strength of colonies randomly as selected by the grower.
2. To leave the bees in the crop for a period necessary for effective pollination estimated to be apporximately ________ days and with a maximum period of ________ days, after which time the bees will be removed or a new AGREEMENT negotiated.
3. To ensure that colonies are properly located and will remain in good condition while pollinating the crop.

Dated ________________, 19______, __________________ (City and State)

______________________________ (L.S.)
(Grower)

______________________________ (L.S.)
(Beekeeper)

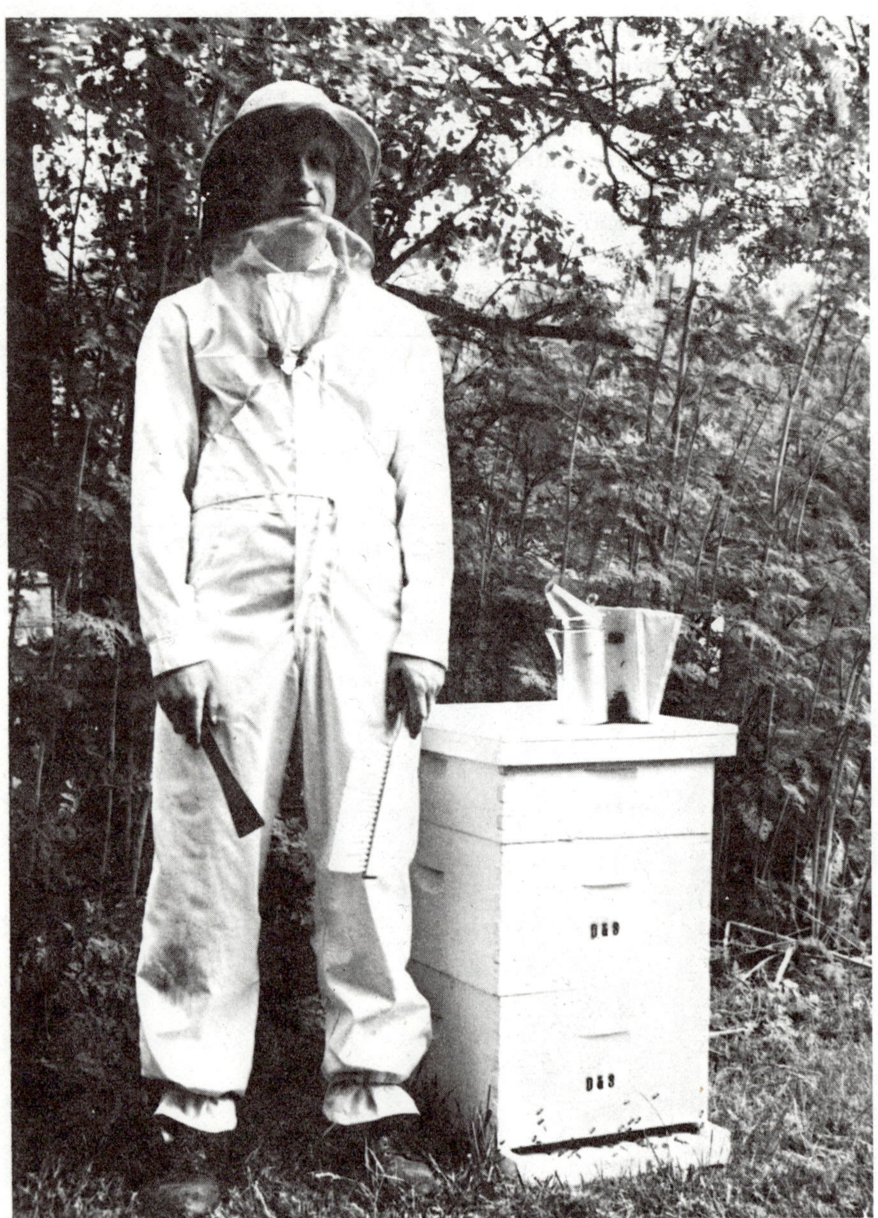

Fig. 11. Friends and neighbors are often naturally curious about beekeeping. Read and keep informed so you can talk about your bees in an informative and down-to-earth manner. Chances are you'll be invited to share your hobby or sideline business with various groups. Use these opportunities to point out the value and importance of bees and beekeeping.

GLOSSARY

GLOSSARY

A

Acarine Disease: Adult disease caused by the microscopic mite Acarapis woodi — not reported in the United States.

Allegation: The assertion, declaration or statement of a party to an action, made in a pleading, setting out what he expects to prove.[1]

Apiary: All colonies, hives, and equipment assembled in one location in the management of the honeybee.

Apiary, Out: Apiary situated away from the home of the beekeeper.

Apiculture: Science and art of maintaining and managing honeybees.

Appeal: Generally regarded as a continuation of the original suit rather than as the inception of a new action.

B

Bee Bread: Pollen that has been collected by the bees, deposited in the comb, and moistened with honey, to be used as food by the bees.

Bee, Bumble: A larger, more hairy and louder buzzer bee than the honeybee.

Bee Communication: Honeybees communicate the direction, distance, and odor of a newly-found source of food to other bees by dances performed on the combs.

Beehive: Man-made box or receptacle for housing a colony of bees.

Beekeeping: See Apiculture.

Beekeeping, Migratory: Moving of colonies of bees from one locality to another during a single season so that advantage can be taken of two or more honey flows.

Bee, Worker: Female bee that has partially developed reproductive organs that carries on most of the routine work of the colony.

Bee Yard: Synonym of apiary.

Bees, Field: Worker bees that collect nectar, pollen, water and propolis in the field.

Bees, Nurse: Young bees of 3-10 days of age, which feed and take care of developing brood.

Bees, Queenless: Stray workers tend to join together to form a cluster, even in the absence of a queen.

Bees, Scout: Worker bees said to locate a home for a swarm of bees. Also, bees that locate a new source of pollen, nectar, propolis, or water.

Bees, Social: Bees living in family groups, usually with a single reproductive female, some males, and a worker caste.

Beeswax: Complex organic compound secreted by eight glands in the ventral portions and on the lower surface of the last four segments of the abdomen of the worker bee.

Bees, Wild: Bees not living in a hive or habitation provided by man.

Bill of Sale: A written agreement, formerly limited to one under seal, by which one person assigns or transfers his right to or interest in goods and personal chattels to another.[2]

Brood: Bees, collectively, in their immature stages.

Brood, Capped: Brood whose cells have been sealed by the adult bees with a porous cover which isolates the immature bees during their nonfeeding larval and pupal periods.

Brood, Chalk: Fungus disease most likely to appear in broods unprotected by the cluster.

Brood, Chilled: Immature bees that have died as a result of exposure to cold.

Brood, Drone: See Drone Brood.

Brood, Purple: A disease of the brood characterized by the purple color of the affected larvae; of obscure etiology, thought to be due to the poisonous nectar or pollen of the titi tree.

Brood, Sealed: See Brood, Capped.

Brood, Spreading: Placing of one or more empty combs between frames of broods to encourage the queen to lay.

Bruising: Any injury by accident, such as the indention of the surface of the comb by pressure of fingers; considered to be damage if it is sufficient to cause leaking.

Burden of Proof: The necessity or duty of affirmatively proving a fact or facts in dispute on an issue raised between the parties in a cause.[3]

C

Cappings: Thin wax coverings of cells full of honey; also the same after they have been sliced from the surface of a comb before extraction.

Cell: Hexagonal unit-compartment of a honey comb. Worker cells are approximately 5 per linear inch, drone cells about 4 per inch.

Cells, Uncapped: Cells, either empty or filled with honey, which are not sealed or capped over by the bees.

Chlordane: Chlorinated hydrocarbon having a formula of $C_{10}H_6Cl_8$, used as an insecticide — highly toxic to bees.

Chuck Comb Honey: See Honey, Chuck Comb.

Cluster: Group of bees which is formed during the winter; also the group which is formed where a swarm settles.

Colony: Aggregate of worker bees, drones, queen, and developing young, living together as a family unit in a hive or other dwelling.

Comb: Waxy structure composed of two layers of cells united at their bases, in which brood is reared and honey and pollen are stored.

Comb, Drone: See Drone Comb.

Comb Foundation: Thin sheets of beeswax with the cell bases of worker cells embossed on both sides, in the same manner as they are produced naturally by honey bees.

Comb Honey: Honey produced and sold in the comb, commonly in thin wooden sections of 4 1/4 X 4 1/4 or 4 X 5 inches.

Combs, Drawn: Combs having the cells built out from the sheet of foundation.

Common Law: As distinguished from Roman Law, modern civil law, canon law, and other systems, the common law is that body of law and juristic theory which was originated, developed, and formulated and is administered in England, and has obtained among most of the states and people of Anglo-Saxon stock.[4]

Conversion: A distinct act of dominion wrongfully exerted over another's personal property in denial of or inconsistent with his title or right thereon.

Corporation: An artificial being, invisible, intangible and existing only in contemplation of law, with a name of its own, under which they can act and contract and sue and be sued.

Cross-Pollination: Transfer of pollen from an anther of one plant to the stigma of a different plant of the same species.

Cut-Comb Honey: Comb honey cut into various sizes, edges drained and the pieces wrapped individually in cellophane or packed in cardboard containers; generally cut from bulk comb honey.

D

Damage: Any injury or defect that materially affects the appearance, edibility, or shipping quality of comb honey, such as:

a) One-third of the combs may have not more than 50 cells of pollen in a comb, provided they are not widely scattered but are on the outside edges of the comb.

b) Presence in comb — section honey, shallow-frame comb honey, wrapped cut-comb honey, or chunk or bulk comb honey packeted in tin or glass, of more than 10% by volume of granulated honey in the uncapped cells, or of more than very small or scattered granules in the capped cells.

c) Presence of any spots of bee excrement on the comb.

Damage, Serious: Any injury or defect that seriously affects the edibility or shipping quality of comb honey. Any spots of bee excrement on the surface of the comb, or in sections attachment of comb to less than 45% of the adjacent area shall be considered serious damage.

DDT: Dichloro-diphenyl-trichloro-ethane. One of the chlorenated hydrocarbon insecticides; usually regarded as less hazardous to bees than arsenicals.

Defendant: The party against whom relief or recovery is sought in an action or suit.[5]

Dextrose: Simple sugar, $C_6H_{12}O_6$,which occurs in honey and on granulation forms the solid base; also called glucose.

Drifting: The drifting of bees from one hive to another because of wind or confusion caused by other circumstances rather common.

Drone: Male honeybee.

Drone Brood: Male brood reared in larger cells than the worker bees.

Drone Comb: Comb having cells measuring about 4 cells per linear inch — have about 18.5 cells per square inch on each of its two sides.

Drone Layer: Unmated laying queen, or one without the ability to lay fertilized eggs.

Drone Trap: Device placed against the hive entrance to prevent drone or queen bees from leaving the hive.

Drumming: Pounding on the sides of the hive to drive bees upward in transferring.

Due Process of Law: The law which hears before it condemns; which proceeds upon inquiry, and renders judgment only after trial — Daniel Webster.[6]

E

Embed: Force wire into comb foundation by heat, pressure, or both, for the purpose of strengthening the resulting comb.

Ethylene Dibromide: Heavy, colorless liquid having the formula $C_2H_4Br_2$; useful in fumigating combs to kill all stages of wax moths.

F

Foulbrood: Infectious bacterial diseases of bees affecting the brood; usually refers to disease caused by *Bacillus larvae* and *Bacillus pluton*.

G

Glucose: See Dextrose.

Grafting: Process used in queen rearing, consisting of the transfer of a worker larva from its cell to an artificial queen-cap.

Granulation: Formation of crystals of dextrose hydrate in honey. The result is a hard or semi-hard composition.

H

Hive: Man-made home for bees.

Hive Body: Wooden box that encloses the frames for holding honey comb.

Hive, Box: Box used to hive bees but lacking movable frames; most states consider it illegal to maintain bees in box hives.

Hive, Decoy: Hive placed to attract stray swarms.

Hive, Leaf: Hive composed of individual frames which, when placed together like the leaves of a book, form a complete hive.

Holes, Dry: Holes in the honey comb larger than a cell, and not next to the wood; they may extend partly or entirely through the comb.

Holes, Through: Holes or passages through the comb from one side of the comb to the other, between the edge of the comb and the section.

Honey: Product made by bees from nectar which they gather from plants (See Honey Chapters).

Honeybee: Any of the bees that store honey in their combs; usually limited to members of the genus *Apis*.

Honey, Bulk Comb: Comb honey produced in shallow frames; usually sold as a complete unit in the frame.

Honey, Candied: Crystallized or granulated honey.

Honey, Chunk Comb: Combination of about 40% comb honey and 60% liquid honey in the same container, usually glass.

Honey, Comb: See Comb Honey.

Honeycomb: See Comb.

Honey, Cut-Comb: See Cut-Comb Honey

Honeydew: Sweet liquid excreted by aphids, leafhoppers, and some scale insects. Liquid collected by bees especially in the absence of a good source of nectar.

Honey Extracted: Honey removed from combs by means of centrifugal force.

Honey Extractor: Machine for the removal of honey from the cells of comb by centrifugal force.

Honey Flow: Period of abundance in the production of honey, during which the bees are able to gather surplus nectar to convert into honey which they store in their combs.

Honey Sac: Enlarged portion of the fore intestine located in the abdomen of the bee and used for carrying nectar, honey or water.

Honey, Surplus: Amount of honey above that needed by bees for their own use.

I

Indictment: An accusation or charge of the commision of an indictable offense, made in writing by a grand jury[7] . . .

J

Joint Venture: An association of persons in a single business enterprise for profit, for which purpose they combine their property, money, effects, skill, and knowledge, without forming a partnership or corporation.[8]

Judgment: The official and authentic decision of a court of justice upon the respective rights and claims of the parties to an action or suit therein litigated and submitted to its determination.[9]

L

Law: The whole body of rules of conduct applied and enforced under the authority of established government in determining that which is proper and should be permitted and that which should be denied, or even penalized.[10]

Levulose: A simple sugar, $C_6H_{12}O_6$, usually the predominant carbohydrate in honey — relatively slow to crystallize and sweeter than either sucrose or dextrose.

Liability: The state of being bound or obligated in law or justice to do, pay or make good something.[11]

Lien: A claim or charge on property for the payment of some debt, obligation or duty.[12]

M

Malice: That state of mind which prompts the intentional doing of a wrongful act without legal justification or excuse.[13]

Mass Feeding: Provisioning of young larval honeybees with quantities of food so that they always have sufficient to feed upon whenever they desire.

Methyl Bromide: Insecticide having the formula, CH_3Br_2, and used for fumigation of combs in storage.

N

Nectar: Sweet liquid secreted by the nectaries of plants; raw source of honey.

Nosema Disease: Abnormal condition of adult bees caused by the presence of *Nosema apis*.

O

Ordinance: The act of the legislative body of a municipal corporation; a local law.

Overstocking: Too many bees for a given location.

P

Parathion: Organic phosphate insecticide of high toxicity to both mammals and honeybees.

Partnership: Where two or more persons agree to carry on any business or adventure together, upon terms of mutual participation in its profits and losses.[14]

PDB: Paradichlorobenzene, $C_6H_4Cl_2$, a white crystalline substance. It is a fumigant and should never be used on combs containing honey as its odor will permeate the cappings and ruin the flavor of the honey.

Per se: By or through itself; simply as such; in its own relations.[15]

Personal Property: Money, goods, and movable chattels.[16]

Plaintiff: The party who complains or sues in a personal action and is so named on the record.[1]

Pleadings: The formal allegations by the parties of their respective claims and defenses, for the judgment of the court.[18]

Police Power: That power in government which restrains individuals from transgressing the rights of others, and restrains them in their conduct so far as is necessary to protect the rights of all.[19]

Pollen: Minute grains formed in the anther of a stamen, which, when transferred to a stigma, give rise to the male reproductive cells; gathered by the bees as a protein food.

Pollen Substitute: Any material such as soybean flour, powdered skim milk, or brewer's yeast, or a mixture of these used in place of pollen to stimulate brood rearing.

Pollen Supplement: Mixture of pollen and pollen substitute used to stimulate brood rearing in periods of pollen shortage.

Pollen Trap: A device for removing pollen loads from the pollen baskets of incoming bees.

Pollination: Transfer of pollen from an anther of a flower to the stigma of that or another flower.

Pollination, Self: Transfer of pollen from anther to stigma of the same flower.

Pollinator: Any agency that transfers pollen from an anther to a stigma.

Pollinizer: Plant that furnished pollen.

Prima Facie: At first sight. In reference to evidence, adequate as it appears, without more.

Propensity, Vicious: A propensity or tendency of animal to do any act which might endanger the safety of persons and property of others in a given situation.[20]

Propolis: Resinous materials collected by bees — used to stick hive parts together and to seal openings.

Propolization: Sealing of hive openings by bees.

Proprietary Function: The function of a municipal corporation in which it acts and contracts for the private advantage of the inhabitants of the city and of the city itself.[21]

Proximate Cause: That which, in a natural and continuous sequence, unbroken by an efficient intervening cause, produces the injury, and without which the result would not have occurred.[22]

Q

Queen: Sexually developed female bee.

Queen Cage: Cage used to confine a queen usually for shipment.

Queen Candy: Mixture of powdered sugar and invert sugar used in queen cages.

Queen, Fertile: Queen that has mated with a drone and has a supply of spermatozoa in her spermatheca.

Queen, Tested: A queen that has mated with a drone of her own race and proved by the brood pattern to be a good layer.

Queen Trap: See Drone Trap.

Queen, Untested: A fertile, laying queen that has not been observed long enough to determine whether she mated with a drone of her own race.

Queen, Virgin: Unmated queen.

R

Real Property: Such things as are permanent, fixed, and immovable; lands, tenements, and hereditaments of all kinds, which are not annexed to the person or cannot be moved from the place in which they subsist.

Repellants: To be effective a bee repellent must be strong enough to overcome the natural attractiveness to prevent honey bees from foraging plants treated with toxic pesticide, yet it must not injure any part of the plant or harm the operator applying the poison. Nicotine sulphate, creosote, carbolic acid, lime sulfur, naphthalene, and more recently, benzaldehyde and propionic anhydride, commonly mentioned as bee repellents, have only limited use because the efficiency of each depends upon its volatility.[23]

Robbing: Stealing of nectar or honey by bees from an alien colony.

Rocking: A movement which may serve as a mechanical cleaning process by which bees scrape and polish the surfaces of the hive.

Royal Jelly: The highly nitrogenous food secreted by the pharyngeal glands of worker bees.

S

Sacbrood: Disease of the brood caused by a virus.

Septicemia: Disease of adult bees caused by *Bacillus apisepticus*.

Skep: Beehive usually made of twisted straw.

Slumgum: Refuse from the melted comb after rendering.

Smoker: Device in which burlap, wood shavings, or other materials are slowly burned to furnish smoke which is used to subdue bees.

Spring Dwindling: Condition in which a colony gradually loses strength; due mainly to a death rate more rapid than the replacement of the working force.

Statute: The written will of the legislature, solemnly expressed according to the forms necessary to constitute it the law of the state.[24]

Stonebrood: A minor disease caused by the fungus *Aspergillas flavus*.

Sucrose: A disaccharide with the composition of $C_{12}H_{22}O_{11}$; intergral part of nectar.

Super: Hive body used for the storage of surplus honey.

Supersede: Replace mother queen by daughter in the same hive.

Swarm: Aggregate of worker bees, drones, and usually the old queen that leaves the parent colony to establish a new colony.

Swarm, Hunger: Desertion of the hive by the bees as a result of a food absence.

Swarming Out: Abandonment of a hive or other home by a colony of bees as a result of some unfavorable condition.

Swarm, Natural: Portion of a colony that issues spontaneously from the hive or other home to form a new colony.

Swarm, Prime: First swarm to leave a parent colony.

T

TEPP: One of the organic phosphate insecticides used primarily as a contact poison; highly toxic to bees (and humans).

Travel Stain: Darkening of the surface of a comb by materials, usually propolis, that result from the bees walking on its surface.

W

Wanton Act: An act performed with knowledge that injury to another is likely to result from the act and with reckless indifference to such consequences.[25]

Warranty: A covenant against failure of an article for a certain specified purpose or for a certain specific reason.[26]

Wax Glands: Eight glands of the worker bee that secretes beeswax.

Wax, Rendering: Process of melting combs and cappings and refining the wax.

Weeping: Exudation or seepage of honey through the cappings, forming small drops which finally run down the face of the comb. It is usually caused by absorption of moisture from the atmosphere by the honey.

Wilfull: An act done intentionally, knowingly, and purposely, without justifiable excuse, as distinguished from an act done carelessly, thoughtlessly, needlessly, or inadvertently.[27]

Worker, Laying: Worker bee that produces eggs which normally develop into drones.

Writ: A process authorizing or commanding the arrest of a person or the seizure of property, sufficient to satisfy the amount of a judgment against the defendant.[28]

CITATIONS

1) — *Mathews v. Underpinning & Foundation Co.,* 4A.2d.788, 17 N.J. Misc. 79
2) — *Wilson v. Voche,* 172 S.E. 672, 48 Ga.App.173
3) — *Willet v. Rich,* 7 N.E.776, 142 Mass.356
4) — *Lux v. Haggin,* 10 P.674, 69 Cal.255
5) — *Siekmann v. Kern,* 68 So. 128, 136 La.1068
6) — *Dartmouth College v. Woodward,* (US) 4 Wheat 518
7) — *State v. Hamilton,* 56 S.E. 2d.544, 133 W.Va.394
8) — *Brooks v. Brooks,* 208 S.W. 2d. 279, 357 Mo. 343
9) — *People v. Hebel,* 76 P.550, 19 Colo.App.523
10) — *Strother v. Lucas,* (US) 12 Pet.410, 9 L.Ed.1137
11) — *Feil v. City of Coeur d'Alene,* 129 P.643, 23 Idaho 32
12) — *Ross v. Franko,* 40 N.E. 2d.664, 139 Ohio St.395
13) — *State v. Heinz,* 275 N.W.10, 223 Iowa 1241
14) — *Wild v. Davenport,* 7 A.295, 48 N.J.L.129
15) — *Burr v. Winnett Times Publishing Co.,* 258 P.242, 80 Mont.70
16) — *Ralston Steel Car Co. v. Ralston,* 147 N.E. 513, 112 Ohio St.306
17) — *Carmody v. Land,* 21 So.2d.764, 207 La. 625
18) — *Treadgold v. Willard,* 160 P.803, 81 Or. 658
19) — *State v. Dolan,* 92 P.995, 13 Idaho 693
20) — *Hartman v. Aschaffenburg,* LaApp., 12 So. 2d. 282
21) — *Omaha Water Co. v. Omaha,* C.A.8 Neb., 147 F.1
22) — *Lemos v. Madden,* 200 P. 791, 28 Wyo.1
23) — *The Hive and the Honey Bee — Dadant & Sons, Inc.,* Revised Edition, P.688
24) — *Washington v. Dowling,* 109 So.588, 92 Fla.601
25) — *Wunderlich v. Franklin,* C.A.5 Ala., 100 F.2d.164
26) — *Barton v. Dowis,* 285 S.W. 988, 315 Mo.2261
27) — *Lobdell Car Wheel Co. v. Subielski,* 125 A. 462, 2 W.W. Harr.(Del)
28) — *Coples v. State,* 104 P. 493, 3 Okla. Crim. 72

TABLE OF CASES

TABLE OF CASES

INDEX

INDEX